COINCIDENTALISM

COINCIDENTALISM

All is chance

J. Nathan

Tea and Vinyl Publishing

ISBN: 978-1-7394062-3-3

CONTENTS

Introduction

The universe was once completely dark, but then stars ignited. Passing similarly combustive thresholds of chaotic complexity, there grew living beings who developed central nervous systems and brains which became conscious and then self-aware.

This is each of us. We are born, we learn, feel, think, and then die, just like everyone who has gone before. We do not know for sure if there is anything after, but it is fair conjecture that the world will keep progressing into a future we won't know. Most likely we will never know anything again; **we will no longer ever exist**. With only the briefest of awakenings from our eternal slumber we are all going to leave the scene remaining forever ignorant of what will happen to those we loved and leave behind, or how long humanity lasts and what sort of societies the future will deliver.

Game Over seems guaranteed.

The old religious beliefs cannot help us, even if well intended and once upon a time necessary. Metaphorically as myths they are dotted with profound wisdom, brought forth via a connection with the unconscious their creators rarely understood, but when forced upon you as literal truth without any room left for self-thought they are akin to brainwashing cults or tyrannical ideology.

Science cannot help us either. It may make life easier via technology and enable us greater freedom for which we must surely be grateful, but even if it eventually provides absolute safety and perfect health in a world of maximum opportunity for eternity, stuck in a metaphysical bubble as we are no equation or machine will ever satisfactorily answer our inherent existential quest to truly know what and who and where we are.

Describing the bubble, and conjecturing as to what if anything may lie beyond, is the purpose of philosophy, which is, or should be, akin to rational religion.

The only way to sum up your own belief or lack of belief is via philosophy, with anyone who is rational accepting it is based on best guess work and that we can't be sure of anything.

Before my own attempt to articulate the nature of our existence is made however, and with what could be described as 'clarifying the obvious,' the problem of the 'semantic knot' must be mentioned. That is when human beings debate a subject, although their differences of opinion can be exactly that, they can also be due to the inherent subjectivity of language.

For example, when trying to define or conceptualise consciousness, if one person believes 'all living beings are conscious,' and another that 'only living beings with brains are conscious,' and a third that 'only humans are conscious,' they may think they all disagree on what consciousness is, when actually all they really need is three different words, for three different concepts, which they could then all agree on.

Of course they may all agree on what consciousness 'is,' and for instance concur for anything to be conscious it must be able 'to be awake,' yet still hold their own conflicting beliefs as to what it means to be awake, and so therefore be clear on their disagreement.

With conversations attempting to define complex subjects like 'consciousness,' the challenge usually involves a mix of both semantic and genuine disagreement, with half the struggle for clarity being to know what you disagree on.

With that in mind, the following text should be taken as a collection of attempted semantic clarification, and not claims of fact. After all, although being aligned with society semantically is of course generally ideal, such that we all use the same words in the same context and maximise meaningful communication, what really matters is the communal aim of an accurately shared concept-to-language pairing between those conversing, which can then be agreed upon or not. In this way although we know nothing we can at least be sure of where there actually is conflict of opinion.

It should also be noted everything within the following text may be completely 'wrong'. It is a work of philosophy after all, not science, partially due to the author's limited scientific and mathematical training or ability, and partially due to the limit of science and mathematics. It is simply an attempt to share ideas.

We have no idea where we are, and all human thought could be completely deluded, but it shouldn't stop any one of us from considering and attempting to articulate an answer to the greatest mystery of our existence; the nature of our being.

Humans are born to philosophise, as much as reproduce and temporarily survive.

CHAPTER ONE

A Precarious Life

In order to consider the nature of our being, it makes sense to consider the nature of life. There is no clear cut definition of what life is, even if it is obvious that an animal is alive, but a rock is not. At one extreme is the view that everything is chemistry, with nothing fundamentally different between that which is living and that which isn't, and at the other is the idea that there is an individual soul, or a life-force, which very much distinguishes that which is alive from that which is not.

Here in this text, life is considered as both just chemistry, as well as distinct enough to be something separate, as is common in orthodox scientific theory. It's significant separation from the rest of the environment is essentially due to its complexity, and nothing more, for despite the search for some sort of life-giving spark or spirit, none have been found, and there is nothing special about the chemistry of life in itself. The extra level of complexity could most simply be tied to the presence of DNA, although this would define a virus as alive which most who consider such things don't want to do. There are therefore generally several requisites to be a life-form, such as having a body, which a virus doesn't have.

Although the nature of the virus and it's half-life status are fascinating, like a program that needs a host, the aspect of being alive which is most important for us philosophically, as well as psychologically, is that it's essential drive for survival is centred around its 'self,' which is founded on its separation from its environment, due to its relative complexity. It is of course true that in the detail everything is connected to everything else in the universe, and so it is in a way true that when looking deeply there is 'no self' as the wise ascetics proclaimed, with nothing completely isolated, but it is also true there is a distinction between a living being and its environment that can't be denied.

From the first living cell with its membrane, which was also a filter, there was a division between what is outside and not me, and what is inside and a part of me.

This division of matter into organic and non-organic matter is due to the different levels of complexity, but equally significant and indeed vital to the evolution of life is the fact that the essence of the complexity is a program to maintain itself, or to survive. This instinct is what makes it seem as if there is a special life force, or a thirst to be alive that inhabits all living matter, but that is like saying a ball has the desire to roll down a hill.

Thanks to what we perceive as an explosion at the beginning of time, our expanding universe is like a river of flowing energy. Life is powered by this river, or is in fact a part of this river, as is everything, but rather than just being random eddies of various complexity and stability formed on its bank, as non-living matter is, life has formed a dam which steadies the flow storing it up into a reservoir as if making a battery. This extra level of complexity is what defines life and makes it *stand out,* despite it also being natural, and not as unsurprisingly assumed, constructed.

All complexity is resisting the flow of the stream, even just standard eddies, which is why all matter is inherently unstable, with some more or less stable in our time frame, and it is the mixing of the relative stability in matter that the program of life is built upon, formed with fine tuned integration.

In this way there can be no life without the precipice of death. It's state of separation is built upon the continuous process of requiring a dam to be constantly reinforced so as to steady the flow of energy, whilst the forces of nature, or statistical probability, drive to destroy it. Every moment it survives the odds of it surviving further decrease, as if we are walking on a tightrope that gets longer with every step.

Due to the inherent battle of life with its own complexity, which just by surviving is becoming more complex, and thus always increasing its odds of disintegration or death, immortality was never very likely. Immortality, the inbuilt goal of life, is in fact probably an impossibility, with every extra minute harder and harder to claim, like mass trying to catch up with light. Even living long is unlikely, and living longer always increasingly unlikely, which is why individuals die and reproduce. Presumably at the time of the first life on Earth after many singular forms existed and were unable to sustain their more complex state, falling off their increasingly wobbly and slack ropes, one of the living cells happened to successfully divide, increasing the odds of some life surviving. Initially these offspring, being duplicates, would have been the same age as their parent, and only improved their chances of continued existence as a species by their multiplication, but ultimately this process developed into the system of reproduction, where the new life-form was essentially young, or in an earlier simpler state to the parent, with a newly tightened rope, even if like everything else they were born on the banks of complexity and growing older by the second. This modified the explosion of life into a conveyer belt of generations, bringing the opportunity to splice in change to the process, with inherited random mutation possible between generations, especially with the advent of sexual reproduction and its genetic shuffle, which when they happened to improve survivability of the individual, gave rise to the diversity of life. The competition between different individuals as to who could live the longest, intensified into a battle for resources with the aim to breed. With life itself a primary resource, each became both predator and prey.

Different species now fight for their position in the ecological web, as well as individuals within species, despite all life being offshoots of the same first reproducing life-form, and part of one system.

To be alive requires a constant battle for survival, not just from competition with other life, but because it is by its very existence walking on a homeostatic knife edge, which is akin to maintaining the balance of an intricate equilibrium.

This is where we come from, and the perch on which our individual self clings, whilst every day that we age, we increase our odds of falling.

CHAPTER TWO

The Conscious Brain

Since time immemorial humans have felt their self divided into body and mind, and we tend to think of our mind or our personality as the centre of our being, with consciousness commanding from the summit of the self, rather than our physical form, even if the one is dependent on the other as is so clearly demonstrated with the unfortunate effects of brain injury.

We also tend to think of the mind as somehow ethereal, existing as brainwaves or inhabiting the body like a soul, when it makes most sense that all mind is founded on matter, and as there is no life-force, so there is no separate mental realm that does not require a brain.

'Consciousness,' which most would agree requires a living being, but lots of people have vastly different ideas on, will here be defined as dependent on having a brain, such that for example plants are not conscious, and do not have consciousness.

There are of course also other living beings which do not have a brain, besides plants, such as bacteria, snails or jellyfish; and some which have neuron's in their body but no specific cluster of them that can be called a brain; and insects with a bit of a proto-brain and nervous system such that the requirement 'to have a brain' is not clearly cut, but all those animals that obviously have a brain, such as mammals and reptiles and birds, are by this definition 'conscious,' and have 'consciousness'. They can be awake or asleep.

Before the brain and consciousness, all the responses of an organism were made exclusively by instinct, the same as they are with bacteria and insects today, but animals which have a brain and are conscious also have feelings and memories to guide how they react.

A feeling is a singular overriding output of the brain from analysis of multiple senses all over the body.

A memory is ultimately a repeated response such that chemicals and genes and instincts are themselves 'memories,' but the brain and its ability to respond with feelings brought about memories as we more commonly use the term. With a brain specific situations can be remembered which then trigger behaviours learnt during the individuals life time, which can be guided by feelings or desires as well as instincts, becoming instinctive themselves.

In this sense cats and dogs have feelings and memories and consciousness, but jellyfish and snails do not. As for insects, although with their 'proto-brain' and nervous system they may have some degree of consciousness, it seems likely to be very minimal.

CHAPTER THREE

Self-Awareness

A very significant development to consciousness and a uniquely defining feature for humanity is what will be termed here 'self-awareness,' or just 'awareness'. Although many animals are conscious and have feelings and even dreams, before the first moment of self-awareness no consciousness had ever registered itself. At some point, somewhere and within someone or some primates, this changed. In the brain of an unknown hominid, there was for some inspired individual of the past, the first ever awareness of time, and the idea that there was a yesterday and will probably be a tomorrow. There was a sense of self or of 'I,' and of the other or of 'you'. There was a distinction between 'me' and 'everything else'. The metaphoric mirror in the mind made its first reflection and the universe had its first thought.

Self-awareness can be defined as the ability to shift the primary focus of consciousness onto a memory, which brought about the ability to remember a memory, and in turn experience thoughts and emotions.

'Thoughts' are essentially a mix of integrated memories, which have been summarised into a symbolic word or image.

'Emotions' on the other hand are essentially feelings, but

more complex, because they are integrated with memories and thoughts and other feelings.

Self-awareness required the primary focus of consciousness to switch from the immediate present to the past, and listen to its echo, and it took a little courage, or some faith in the safety of an auto-pilot. To look over your shoulder was frightening, like driving down a road at speed and casting your eyes onto the rear view mirror, instead of always forward not even blinking.

To see yourself truly, as a living animal surrounded by threat, be it from predators or just thirst and hunger and the constant need that life demands, must have been terrifying.

This was the birth of mind, which opened up on a horror show.

No other animals on Earth have a mind.

No other animals have a sense of self.

No other animals know they are alive.

Though they have feelings, and memories, they are not aware of them and so do not experience emotions. Though a dog can have dreams and get anxious or excited, and many animals can feel pain, they do not know they are feeling pain, and so in a sense do not experience pain.

Since awareness, decisions can be made emotionally, and 'consciously'. 'Conscious decisions' require awareness. Even though an animal such as a cat has consciousness, it can't make 'conscious decisions'.

The ability to remember a memory brought great advantages, with apparently prophetical abilities for those who could control it, but it also brought dangers. It brought the knowledge of death, which was everywhere and around every corner, with the possibility of overwhelming fear or terror and panic. In the early days of awareness, the mix of imagination and the terrible conditions must have been bordering insanity, and it is not surprising that there was war, and butchery, and ancestor worship, and all sorts of madness like cannibalism, which although shocking today was then popularly believed to be a way to absorb someone else's power, which understandably made sense at the time, if with a rather gruesome logic.

For the first generations of mind, it was like they were constantly on hallucinogens, or living in a world of children with no parents, and all very 'Lord of the Flies'.

For this reason, and very important for all future descendants psychologically, was the presumably soon to follow ability of 'repression,' which acted like a sanity-valve. That is, just as non-aware animals can be influenced by memories whilst remaining unaware of them, the human brain learnt to utilise its internal conscious reflections such that they themselves became instinctive. Human beings can think emotionally or with internal words, which requires awareness, whilst remaining unaware of their thoughts.

Adult humans can make conscious decisions they are unaware of, as they can speak without having to consider every word, having learnt to talk automatically.

These unaware thoughts and emotions or calculations which utilise consciousness and awareness are called unconscious, or subconscious, and alongside instincts they can have profound influences on human behaviour, subtly channeling dreams and aspirations, or not so subtly erupting into consciousness, whether the person be aware of them or not.

It is very rare for a human being to use their full capabilities and be aware of what they are doing, which takes a form of mindfulness few have time or inclination for. We use our instincts and emotions to help guide us for good reason: our time is finite. We want to get on, and get things done, in order to survive and prosper. We don't want to sit around over-thinking!

Since the jump to awareness, human beings have learnt to contemplate their environment and themselves in ever increasing detail, with an explosion of information gathering. As we are all aware this has risen to new significance with modern media, and there is a staggering diversity of opinion and analysis available to a modern human compared to an ancient one. Once again we are having to learn how to turn it off.

CHAPTER FOUR

Rational Training

An important knock on effect of the ability for humans to be aware is the ability for humans to be rational. Rationality can be defined as 'the ability to utilise our awareness for critical reasoning'.

Critical reasoning can be defined as 'the resolution to question any idea suggested as a 'fact' with as much contrary speculation as possible, such that only the most likely 'facts' survive'. It is a simple enough technique, which acts like an honesty filter on all information processed, minimising any personal subjective distortion or bias, and benefitting from as varied and numerous a sample of opinions or observations as possible.

Both rationality and critical reasoning were given a catalytic boost by the invention of writing, which is essentially an external, visible, and much more stable memory device, compared to the human mind.

The ability to be rational, as well as the ability to read and write, stand out as markers within conscious progression, almost as much as being aware. However unlike awareness, which comes naturally to every individual, emerging behaviourally in infancy such that it is now an instinct for our species, rationality, like reading and writing, generally requires training. Rationality is a skill which must be learnt.

Without time taken for reflection, be it due to lack of training or because of an immediate situation, our feelings and emotions, conscious and unconscious, can always get the upper hand, and take over our decision making, especially when we feel threatened.

Responses based on feelings or emotions are more likely to be irrational. They utilise stereotypical summaries and are formed impulsively, jumping at the guess which feels right, without taking the time to reflect on all the information calmly.

The vast majority of our decisions, conclusions and responses are made this way. Human beings are rarely rational in the moment, even when they think they are.

One of the most profound products of our ability to be rational is of course science. It is the progress of science, and the technological development alongside it, which has enabled us to unearth so much more information.

Another is philosophy. Philosophy is the attempt to honestly answer the questions which science can't, largely to do with metaphysics, and morality.

Two of the oldest metaphysical questions that are yet to be rationally answered are '*What are we*?' and '*Where are we*?', both of which have had countless often very wonderfully imaginative answers, many of which include irrational descriptions of miracles and divine beings.

For the modern day rationalist, able to easily survey and peruse the great quantity of different recorded opinions from the past, it is of course very unlikely that any one of these ancestral ideas are literally true. It is very unlikely that there are gods, or that there is a 'God,' at least in terms of being a man with a white beard sat upon a cloud watching us and monitoring our behaviour. Nor is it likely there are any ghosts, or aliens, or werewolves either, although their prevalence in mainstream media might make you think otherwise.[1]

---

[1] The adoration and belief in gods, who once came to Earth and may do again, is akin to believing in super-hero cartoons; and the adoration of an external and divine Mother and/or Father is an expression of a longing for the delusion of perpetual childhood, or a return to The Garden of Innocence/Ignorance. The Old Man with a beard is basically an infants fairytale, yet millions if not billions of adults alive today believe these sort of stories to be absolutely true; a clear demonstration of the hold irrational beliefs still have on humanity.

This takes nothing away from the metaphoric value of ancient beliefs, (be they mythological, religious, or philosophical) some of which remain profound. Indeed to criticise or ridicule a scripture or myth literally in modern context is as naive as to believe in it literally in modern context.

An obvious example of where metaphoric value remains but literal belief does not, is the myth of 'The Fall,' (which according to The Bible is the moment when humankind were kicked out of The Garden of Eden because they became knowledgeable, thanks to Eve eating forbidden fruit, tempted by the 'Devil' disguised as a serpent). This has been taken as factual by millions of human beings, with some otherwise rational and educated individuals even searching the globe for the actual garden where they thought it happened, when the greatest miracle of the story is that it fits perfectly as a metaphor which describes dawning self-awareness. Psychologically it expresses the horror of our 'knowledge,' because ever since then, and unlike other animals, we are able to comprehend our mortality. The secret is out the bag, we are all going to die, and it would be much easier if we could go back to where we were before (similar to the myth of Pandora's Box of course, in which a woman is blamed for bringing the 'evils' into the world). It is incredibly 'insightful,' even if The Garden of Eden is actually a jungle and more hell than paradise - and knowledge is our possible 'saviour' not a curse of the Devil, and it is completely unfair if quite typical of the past that men blame women for their ability to feel pain and die. Yet those that wrote it were not necessarily aware

Prayers also don't work, even if like chants, or crystals, or rituals, or astrology, they may help you feel better, or decisive, and so have circumstantial or placebo effect.

The reasons lots of people still cling to outdated beliefs, as well as superstitions, or ritualised good luck procedures, and even early philosophical conjectures (such that there are four elements earth, wind, fire and water), are obviously psychological.

Belief systems are means by which we combat our inherent default state of anxiety, or existential trauma. They are like scaffolding and are vital for our psyche and confidence to grow, and as necessary for the human mind as a skeleton is for the body. Without some form of belief, if only in tomorrow, or the love of your family, a human would be unable to speak, unable to move, and would ultimately die.

Many beliefs are formed when we are young, and become like our language engrained so deeply in our consciousness they are almost instinctive. Learning how to talk we believe in our words, and our grasp of the meaning of the words. We form maps which correlate the world to our imagination, and the more faith we have in them, the more permanent the ink with which they are written.

Once beliefs are formed, it can be difficult to change them, and take away their support. It can even be difficult simply to question them, because of the way we will have emotionally invested in our decision to believe them, and the strength they have given us. Facing the fact your world view is just a fantastical opinion can be challenging to one's sense of well being. If a decision has been difficult, we do not want to revisit it. It can benefit the chances of our survival if we act

---

themselves of its insight, and their own insecurities, or their own intimate connection with their own unconscious, which we no longer experience as easily when adults, it being somewhat sacrificed by our rational progression.

more than think, and are not always second guessing. Impulsive decision making, not least the decision to breed, has been selected as beneficial to survival by the process of evolution, and the same is true for those who have a relatively well structured set of core beliefs.

New ideas, especially about the big metaphysical questions, are more likely to be agreed with and spread across a population if they increase a sense of confidence and meaning in existence. Science is an incredibly powerful tool, but beyond the here and now it is not much of a belief system. It has not yet conquered death, and being rational, unlike other belief systems, cannot claim to have done so. Though it has extended life significantly, it has not yet discovered the Fountain of Immortality, which like other belief systems, is ultimately what it searches for.

In this sense it has so far 'failed'.

Some prominent scientists have said when they discover a Theory of Everything, and scientific equations can describe how The Universe came to be and everything that it is, there will be no need for philosophy. Einstein, the grand wizard of science, or its most famous modern day prophet who took the mantle from Newton, died still trying to find the Theory of Everything, paper and pen in hand as he passed away, with his brain soon dissected by his followers hoping to find the source of his power. But even if physicists discover a mathematical solution which describes the nature of our universe to all possible detail, and extend the human life to thousands of years, a formula can never provide, or take away, meaning. When everything is described, if that is even possible, there will still be a need for philosophy.

CHAPTER FIVE

In Defence of Honesty

Rationally we know we will follow all those before us who have climbed over the top one after another into oblivion, with death truly the only certainty of every life. Rationally our significance is tiny, if we can be thought to have any at all.

Even though most believe there is difference between right and wrong, this too, if you are rational, is agreed to be only subjective. Aside from any conscience, to have killed or raped and been a thief of the very worst kind, will it seems have no self consequence if the 'sins' are undetected whilst alive. Tyrants that died peacefully in their sleep will apparently have no further repercussions. Rationally there is no divine justice.

Rationally, life is a battle for power.

Fighting for power, be it through wealth, politics or personality, is the game of life, and the only rules are the ones we agree on. There is no objective morality. There is no reason to be moral except for self-promotion.

Evolution has as we all know crafted the instincts in the fight for power, as well as weapons like tooth and claw.

Power in social groups is enhanced by relationships

between members, teaming together, forming networks through which it can more easily flow, like electricity and energy.

The brain is basically another weapon, sprouted by one of nature's developmental feedbacks, as with all the other predator-prey arms races. Mental interactions are effectively just like physical ones, with every thought a material response in the brain, where the articulate can dominate the stupid, or the scared, and push their points of view. People construct belief systems therefore, as individuals or groups, and try to convince others of their validity, not only in order to normalise their reality and maintain a state of confident action, (to combat their own existential trauma), but also to gain influence and power. Every conversation, even a friendly one, is a battle of power. Just to speak at all is to claim power over yourself, and combat your fearful silence, the same as when you speak your first word.

It is difficult with these facts for anyone that is rational to have any faith in anything but a lack lustre belief in their own back yard. To put yourself and those in your power network forward, and agree to compromise only when it suits you. This is pretty much how most people live, naturally heavily bias to our own point of view and purpose. It seems there is nothing beyond the horizon, and nothing objective to believe in. Being a racist, a nationalist, a humanist, or even a vitalist, (and including all life in your tribe), is all very well; there is plenty of subjective meaning whose power is in its very creation, but there is, most definitely, if you follow rational reason, highly unlikely to be anything beyond.

This is what we are taught today, whether explicitly or not. This is the experience of the modern human.

It is however false.

There may not ultimately be anything completely objective, but there is likely a cosmic morality, and there is likely a higher truth beyond our universe and our short mortal life, which any civilised rational and self-aware species would agree with.

Honest consciousness is an integral constituent of any true civilisation, which on Earth is yet to be founded. This is because despite there being no objective 'truth,' the only nugget of 'truth' we have is that objectivity is maximised when being as honest as can be.

Though it may not be possible to believe in any truth, it is possible to believe in honesty; in wanting to not lie. At least regarding anything serious, and especially regarding your own self.

For example, an advanced civilisation of any aware life form will understand that every individual with awareness maximises the potential of themselves and all individuals and the societies they form as groups with a maximisation of freedom of belief and information, alongside an understanding of rationality and critical thinking.

Tyranny is for primitive societies, and those that wish to be tyrants have infantile minds, like the insecurity of a bully, whose own fear remains buried inside. They are not stable people, because they are not honest people.

This is obvious but it is extremely common for people to feel sure their own view is the true one, regardless of the subject. It is rare in general conversation people take a stance of wanting to learn. This is not just in adult to child relations, but adult to adult, and friend to friend.

In this way we are all actually tyrants. We all tend to want people to share our view, because it makes us feel safer.

To force someone to accept an opinion is not as genuine as having someone voluntarily and openly agree with you, whilst all are aware of available counter arguments. If you are rational, if you are honest, you have to agree with this.

Being able to access all possible information, and jostle

different view points, whilst also being able to rationally make and regularly revisit a decision, is essential to maximising understanding. This is when human individuals are being as honest as they can be, which will lead to a society which is most aligned with our existence.

Even though children must be able to lie to themselves, and be reassured incorrectly by their parents everything will be alright, or that they will always be there, such that they may develop a solid self-identity, and adults may want to lie to themselves on a regular basis in order to have fun, the ultimate aim of education should be to teach honesty.

This does not mean we should aim to eradicate emotion, or irrational arguments, or even lies, which all have their place. In the future humans will be able to modify their genomes and pacify their responses. They will be able to choose to have less emotional offspring that are more inclined to being rational, but most do not want humanity to become robots.

Rationality is the rudder, it is not the drive, or the destination. Emotions are profound, powering us along like a sail, so it would be a shame to lose them. They are part of the conscious landscape, which like the ecological one would be worse off with reduced variety.

CHAPTER SIX

Reality Exposure

Originally humans were very confused at how some of the patterns observed in nature were beyond their influence, whilst others could be affected by their own actions, and so they imagined there were divine beings, which were nearly always the aspects of nature they couldn't control personified. A battle line was drawn between the will of the individual, or the tribe, and the will of the gods.

Gradually however, rationality began to dominate, supported by technology and observation. It was discovered that there are laws of the universe, written as mathematical equations, or 'functions,' which are the same as computer programs, where various methods of arithmetic combine variables into new products depending on circumstance. These core functions or equations crafted existence via their physical representatives physics and chemistry, which together drive biology, and could be used to accurately predict many very specific aspects of the future. Existence in its basics was simple, even if in its detail there was mind boggling complexity, but it was believed eventually with enough reason everything would be predicted, without any need for gods, only equations. The world was a machine, in what was, during the Industrial Revolution, the age of machines. Even animals were machines, of a biological kind, although humanity of course remained in a class of their own. Time moved at a steady rate, and the size of the universe was constant, with only humans as 'free agents' moving around in the space inside it.

Yet it soon became apparent that there was a lot of fuzziness, seeping in from the edges. Human ignorance was, as had long been philosophised, an integral part of the system. Though knowledge and rationality could be maximised, they could never be fully realised. Just as ants will never understand the Earth and the planets, we will never understand that which is more complex than we are, and lies beyond our universe. A lot of scientists still don't believe this. They cling to the hope one day we will understand everything, and in this sense theirs is another irrational faith.

The challenges to the recently cemented scientific model of existence, came from two main prongs of attack. One was 'relativity,' and the quirky behaviour of time and space as mass approaches top speed, which is of course light speed, and is like a limit on how big anything can be. The other was 'quantumness,' and the quirky behaviour of time and space as mass approaches no speed, which is like a limit on how small anything can be.

The walls at the edge, and the holes in our centre, have now been recorded and observed. They are still there, surrounding us inside and out, apparently unconquerable aspects of our existence. They can be described simply as the limits of our universe, and our knowledge, for despite our confident every day relation with them, space and time both remain conceptual riddles, which have never been understood. Nobody can conceive a beginning or end of time or space, yet it is equally baffling that they may go on forever.

CHAPTER SEVEN

Time Bending Relativity

As far as relativity goes, it was realised that instead of everyone and everything travelling through time at the same steady rate, as was assumed, everything travels through time at varying rates, depending on their rate of acceleration, or their mass.

However the concept of a varying rate through time is already misleading.

Everything travels 'through' time at a steady rate, from their own perspective, because this is what time is, or what time is constructed by; 'a continuous perspective,' but everything travels through time at different rates, compared to other perspectives.

Whenever anything with mass accelerates, or is accelerated, their rate through time compared to anything else that isn't accelerated, slows down.

Whenever anything increases in mass, their rate through time compared to what it was, also slows down, the same as being accelerated.

Being accelerated, or increasing in mass, are effectively the same thing. If something is accelerated, it can be thought of as increasing in mass. If something increases in mass, it can be thought of as accelerating.

Both can be thought of as having a higher rate of energy flow. Eating some food, gaining mass and energy, is acceleration, even if you stay sitting at the table. There is just that little bit more energy going through your cells, some of which will be vibrating a little quicker, such that you are 'moving' a little faster, and your time a little slower.

The bizarre fact that acceleration can alter the rate of time is well beyond reasonable doubt, and has been succinctly demonstrated by having very accurate atomic clocks put on airplanes, which when they return have not ticked as many times as any identical synchronised counterpart clocks left on Earth. Their 'rate through time' has slowed. The same would have happened to the molecular structure of the airplane itself, which was also accelerated, as well as any passengers on board.

The difference can be very accurately predicted by equations, formulated most famously by Einstein, which also take into account the smaller countering increasing rate through time by the fact the airplane is in the sky and further from Earth, and therefore less affected by Earths mass. Although time runs slower for the clocks because they have accelerated to faster speeds compared to the surface of the Earth (though still very near to our speed compared to light speed), they also run slightly faster the further they are from Earth, where the gravitational field is less dense.

On the surface of the Earth we have to put in more effort to not be pulled to its centre, the hard work being done by the ground and our muscles, such that we are 'accelerating' even when we are lying down sleeping. On the surface of a Black Hole, if it were possible to exist there, a single unit of mass would have to travel at light speed, or maximum continuous acceleration, just to remain 'standing still'.

The effect of time slowing down because of acceleration, as

well as speeding up because of less dense space, are also demonstrated by the necessity for satellites to have their onboard clocks artificially synchronised, depending exactly on their speed and orbit, so that they run on Earth Mean Time as it were, the average speed of time on Earth. Technically however if we wanted to be very accurate, we would do this for every unit of mass, not just clocks on satellites. When we wave our hand it travels 'through time' slower than our body, and our rate through time varies when we get in a car or travel from the summit of a mountain to the sea, but the differences are so tiny, and have no relevance to our normal lives, so can be ignored or averaged out, as by the watch on our wrist, which is perfectly acceptable being able to measure in seconds. If our watch could measure to the smallest units of time used by our universe, and every unit of mass had its own clock, you would be able to notice the watch on your waving hand going slower than the one on the wrist which remains by your side. You would notice every unit of mass balancing its rate through space with its rate through time, such that its rate through 'spacetime' remains the same.

In an often used thought experiment, the airplane is instead a spaceship and able to go to the sun and back, or better a star some lightyears away, and accelerated up to approach closer to light speed, and the clocks become human twins with one onboard and one left behind, with the result that when the travelling twin comes back they could be ten or more years younger than the one which stayed home growing older faster than they have. If the ship had gone far and fast enough, they could return to an Earth thousands of years in the future, with the one twin barely aged and the other long dead.

This remarkable fact has frequently been interpreted as supporting the idea that it's possible to make a time machine, if one that can only go forwards, but it is very important to remember that the twin onboard, and any other passengers, don't see or feel their own time slow down. Except for the fact the accelerated mass or person would feel heavier as they accelerated, they would not feel as if their own journey through time had changed rate. They don't see their own watch or the clock on the spaceship slowing down. They don't start moving in slow motion, whilst carrying on thinking at normal speed, or some other such weirdness. To them it seems as if they just sped up in space, with no effect on their own time, whilst everyone and everything outside their spaceship accelerated their speed through time. By accelerating in space, they have in effect, fast forwarded their environment!

If they could keep an eye on everyone, say by going faster and faster but maintaining the same orbit round Earth, rather than flying away from it to a distant star, looking out a window down onto the planet, for them it would be as if everyone on Earth started speeding up, which would be spinning faster as well as getting smaller, with all the shrinking people rushing around much quicker, whilst their onboard clock and movements and heartbeat carried on ticking at normal rate. When they slowed down again, the Earth would return to its normal spinning rate and size, with all the people on the Earth going back to their normal walking pace, having been accelerated in time.

Of course, those that form part of their environment don't feel their time speed up. The twin left on Earth who is now much older compared to the twin who left, will have experienced more time, and been able to do more things, than the space traveller who is still younger.

Those on Earth, if they looked up at the orbiting spacecraft going faster and faster, would see those on board, and its own molecular structure, moving slower and slower, whilst getting longer and longer, as well as thinner and thinner, stretching across the sky. If the spaceship had approached light speed, and been whizzing across the sky several times a second, the passengers inside, if they somehow hadn't been squashed and stretched to death, would be barely moving, their clocks hardly ticking, frozen not moving in time, whilst those on Earth carried on normally at normal speeds.

There would in other words be different versions of what had happened.

It seems remarkable we can fast forward everyone and everything else in time by accelerating in space, and is remarkable, except it must be remembered everyone else hasn't fast forwarded in time from their own perspective. From everyone's view of themselves time carries on at the same rate, so it is misleading to say time slows down, or that you can travel into the future. You travel into *a different future* than you would have if you hadn't been accelerated, but it is still your future, approached at the same rate. In fact every time you move, or do anything even as simple as turning left instead of right on the road, you change yourself, and the whole world, if only by a very small amount.

One other important feature of relativity, aside from time dilation, and equally supported by all observation, is not only can nothing travel faster than light speed, but no matter your own speed, light speed always remains constant. This is of course a consequence of time dilation. Because your 'rate through time' relative to everyone else slows down when you accelerate, everyone always measures light speed the same. Even travelling at 99% light speed, a beam of light would leave you at light speed, if you measured how fast it zoomed away from you, whilst those watching you would see you travelling almost as fast as light, and thus almost keeping up with your own beam, which would be very slowly pulling away.

Light speed, seems to somehow be like a linchpin of

cosmology.

Light speed, and time, and space, and mass, all need to be re-evaluated philosophically in light of the revelations formulated at the beginning of the 20th century.

It is incredible that all the philosophers that lived before this age, no matter how well tuned their brains or perceptive their wisdom, had no idea about the relativity of time.

Thanks to people like Einstein, all of us alive today, no matter how coarse our brains or dull our wisdom, have a huge advantage over those who went before when it comes to considering metaphysics.

CHAPTER EIGHT

A Quantum Untangling

As far as 'quantumness' is concerned, the other side of the two pronged attack on the relatively new belief in rationality, the revelation of the quirkiness of the behaviour of particles, particularly small ones, was epitomised by what is known as *The Dual Slit Experiment*. Unfortunately it is a little more difficult to demonstrate compared to putting one of two accurately synchronised clocks on an airplane and having predictable specific difference, although it is equally as baffling and apparently contradictory to common sense.

In the experiment, small particles are fired one after another across a box that is as close to a vacuum as it can be, with high likelihood therefore the particles will not hit any others whilst traversing across it. In this way they are like the smallest units, which have the least likelihood of hitting, or interacting with, any other unit. Because humans can't deal with the smallest particles, their being far too small to spot let alone make experiments with, putting the smallest particles you can control or generate easily in what is as close to a vacuum as possible, acts like a magnifying lens, opening up what is happening all the time at the smallest levels.

There are two slits in the far wall of the box, through which they can escape, and the slits have detectors which can be turned on or off, so it can be noted which exit was used, and beyond them is a sensor wall, which the particle will hit whether they are detected or not.

With the detectors on, or even just one of them on, when particles are shot off one after another (in such a way that they have equal odds of escaping through either exit), they mark out two lines of impact on the wall opposite the slits as makes sense. It is like marbles being shot inaccurately at two letter boxes side by side, with lines of impact craters the far side.

If the detectors are both off however, so that it is impossible to know which slit was taken just before they hit the wall, a different pattern is displayed, which is one of multiple lines not just two. The units still impact the wall one after the other, but this different pattern emerges over time.

It is a pattern easily recognised as it is the same as what happens when a wave hits two slits, with distinct interference patterns with peaks and troughs, like a wave of water hitting two openings in a harbour wall, and merging the other side.

The big difference between a wave of water and this experiment of course, despite the similar result, is that a wave of water is made of many particles all going through the slits at once. In this experiment the particles are shot off one after another, and it can go on for as long as you like, or have large gaps in time between each particle being shot off. As long as nobody or no thing observes or records which exit the particles take before they hit the screen, they will build up a wave pattern as if they are waves, or interacting with each other, despite being sent one after another!

As soon as you turn on a detector, the wave disappears and the original pattern of particles returns. Even more peculiarly, this still happens with particles if you just have one detector and it records nothing. For by doing so you 'know' the particle has gone through the other gate, and this will be enough to 'collapse the wave function,' as the phenomenon is referred to in scientific terminology, making the pattern go back to the original two lines, such that the particles are acting like particles and not a wave. This makes it seem as if human knowledge is perhaps involved, or consciousness, but it is not human knowledge per se. As long as there has been a pathway somewhere in existence by which the information of which slit was taken exists, the wave function collapses, immediately. It doesn't matter if a human has watched the detector or not.

This is still very strange though, and it is understandable why every human that has thought about it has been baffled by it, including Einstein.

It needs to be explained how particles that are travelling through the slits one after another are somehow interacting with each other, and how it is possible for the wave function to collapse instantly just because there is a detector on the other slit, which implies instant communication, when it is 'proven' even information can only travel at the maximum light speed.

It seems as if each particle spreads out 'in space' to form a wave, which then somehow collapses back into a unit 'instantly' when it hits the wall, or spreads out in 'time,' so that even though the particles come after each other, they interact with each other, in one 'stretched out' moment of time.

Both explanations obviously defy logic, that a particle can be in more than one place at once, or more than one time at once, yet there are working quantum computers, as well as plenty of other modern technology, which utilise a quantum wave function and the fact that a single unit can indeed be in more than one 'place' or 'state' at once. As with relativity, the equations that predict the behaviour are utilised in very reliable detail and have to be considered in the construction of many aspects of modern technology in order for the technology to work.

Interestingly the issue with the instant communication, known as 'quantum entanglement,' and also a meticulously demonstrated phenomena, though at first hugely challenging to the light speed limit of relativity, came to be less contentious than at first thought. The slits in the experiment are essentially 'entangled,' and although the instant communication between one and the other of the information 'the particle did not come this way' seems at first analysis to break the light speed barrier, it was found that this communication only ever works for random information. In probability terms, if one slit is 'heads,' the other will be 'tails,' and as with a tossed coin, if one slit is used, the other inherently won't be, but up to the moment of the collapse nobody knows which will be which. Right up to the moment of collapse both slits are still 'spinning' and have even chance of being either heads or tails, but they are 'entangled' because one will always be the opposite of the other, and making one collapse will make the other collapse, instantly, even if at opposite ends of the universe. Random related information can in this way be revealed instantly across space, but meaningful ordered information still has to travel at the much slower light speed.

The only real explanation for both relative time and the wave-particle duality of mass, is what has been called the 'Many Worlds Interpretation,' which is the idea our universe divides at every possible interaction, with the result that there are lots of other universes or 'dimensions' with alternate versions of ourselves and our worlds. If true, our decisions, as well as circumstances, determine not our path in this world, but our route through all the different possible worlds, such that we are all travelling into different worlds at every moment. This seems pretty out there, but whilst theoretically possible, can't be proven one way or the other. Some people explain it by saying all the other worlds don't actually exist. The possibility of other futures is opened up and exposed by quantum experiments, but without them reality collapses very rapidly into our one world, the centre of all fractures of possibility, establishing on its path relative probability and variable levels of certainty.

CHAPTER NINE

Coincidentalism

For some the theory of many worlds is ridiculous, but from the evidence the only real question is whether the others are as real as ours. The idea we live in a multiverse is a much more rational and modern view than the idea ours is the only universe, surrounded by nothingness. It is the same as when humans thought the world was only a few thousand years old, or the Earth was flat, or that the universe was just our solar system. Everything has always been pushed back, getting bigger and bigger.

If the multiverse is real, and there is no reason it is not, then it must be infinite; the biggest we can get. It has no limits.

When you look into space, and imagine what may be beyond our universe, you should not imagine endless nothingness, or darkness, but instead imagine endless everything, or light.

Nothing is the one thing there can't be, for if there was nothing it would not exist. It would be over instantly.

An infinite multiverse does not mean there are not finite forms of existence however. In fact it makes much more sense that whilst 'reality' or the multiverse is infinite, all forms of 'existence' or universe are finite.

If reality is infinite, and there are infinite finite universes, everything that can exist, does exist. It might be possible that some things cannot exist, but they can not be imagined either, so we need not worry about them.

The Multi Worlds Interpretation is popular not just with some scientists, but also science fiction writers, although as with any plot involving time machines (which are absolutely impossible), as soon as there is any communication between worlds, spacial or temporal, it rapidly falls into science fantasy. For if there were any communication between two different 'worlds,' this would make them the same world!

Rather ironically, scientists often make a similar mistake, and without the artistic licence of a fictional writer.

For despite referring to an infinite multiverse, or even an infinite universe, they still usually imagine an infinite number of finite 'beginnings,' or an infinite number of 'initial conditions,' from where there is a clash between randomness and order, which intertwine to create a partially predictable chaotic system.

Despite some whispers of a 'Block Universe,' where future and past and present all exist forever, the majority still believe in ideas like a beginning to time, or that there are chains of cause and effect, when if the multiverse is infinite, there is no reason for there to be cause and effect at all.

This is a shift in perspective that doesn't seem to have caught on, unsurprisingly given how preposterousness it is.

From our perspective it of course seems highly likely that there are cause and effect reactions with forces acting between particles; that when we kick a ball there is impact between our foot and the ball, with energy which is transferred but never added to or reduced, and all of which can be very accurately described.

It seems highly unlikely that there is no cause and effect,

and every predictable pattern now and in the past is just chance, including the structural levels of energy and the way it moves. It seems unlikely that actually there is technically, if not practically, no such force as gravity, and that we are all as likely to start floating up as we are to keep our feet on the ground!

Many people have spent their lives repeating experiments trying to discover the laws of our universe, from a pendulums swing to a planets orbit, and clearly they do work very well. They have discovered extremely accurate predictors, which never fail. Everything we know, from our own physical bodies to our technological advances, rely upon them. Never once has someone started floating, or a pendulum defied its predictable pattern. Yet if reality really is infinite, and the 'Many Worlds Interpretation' is correct, there are no need for forces at all. In fact, because of the apparent nature of information and mathematics, it is more logical that there aren't any.

If everything that can exist does exist, because reality is infinite possible information, absolutely everything can be randomly generated. Every movement, and every decision, could be the product of chance.

This is in fact the most rational conclusion.

The belief that everything is determined by pure random chance means not only is there no choice or 'free will,' but in a strict sense there is not even chemistry, nor any laws of physics, at least in the way commonly understood with 'forces'.

Instead, every single moment or movement (of which there are many zillions every second), every single 'smallest unit of energy' has the possibility to randomly alter its state, whether it be considered to have one of two options (like a coin or a binary information system), or one of many options (like a die or an information system with more than two 'dimensions'), and the random result produces patterns that look like movement or influence, from a relatively fixed perspective.

This idea is well represented by the famous proverb which states that if you have an immortal monkey banging randomly on a keyboard, they will eventually write a word, and then a sentence and ultimately, though it be extremely unlikely, the works of Shakespeare.

Our existence is the same, except the words, or even the letters, can be typed out of sequence. They are eternal possibilities, which create all the various forms of existence by being collected in all possible sequences. All combinations exist. All patterns exist and are eternal. The writings of Shakespeare were randomly generated, as was Shakespeare himself, that is his body and his ideas, and the world in which the writings of Shakespeare, exist.

In some random patterns, 'time and space' and the 'perception' required to generate them, do not emerge at all. They are too random.

In some random patterns, time and space emerge but only in rather jumpy and discontinuous groupings, like an insane universe, or an unedited book.

In some random patterns, the weave is frequent and steady enough for time and space to appear as effectively continuous and without any holes, like a tightly spun cloth, or a well written story. There are random repeating patterns which can give rise to the possibility of a sense of movement through space and time, with a conscious perception such as ours.

The greatest support of this belief is that it explains everything, and is the simplest of rules. There is no need for any other laws or facts. Everything is chance, reality is infinite, and there are infinite different finite sequences that exist, of which each of us are one.

The greatest criticism is that it means the odds of anything continuing to exist, especially something as complex as our universe, or ourselves, is stupendously low.

If everything is random, when someone throws a ball it would have as much chance to appear anywhere in the universe as in the trajectory determined by our laws of physics. The greatest odds would be that it disintegrates. Yet this never happens.

On Earth, from our level of perception, the laws of science are never broken.

The criticism that a world as logical as ours is very unlikely, is of course countered with the idea that if reality really is infinite, somewhere a lucky world that has sustained predictable patterns like ours would exist.

The 'goldilocks effect' does not set the conditions in which there *is* cause and effect, it sets the conditions where there *appears to be* cause and effect.

Another main criticism is that it makes no difference, and can't be proven. Even if the laws of science are not real, and everything is random, and the odds are that our universe will disintegrate in the next moment, we have to believe in physical laws. They have worked for so long it is likely, relatively speaking, they will continue. Our world has been so lucky, and at time of writing continues to be so lucky, that we can make laws to predict what will happen next!

This is true.

We should not give up on science or rationality, which subjectively for us certainly exist, but we should consider the implications philosophically. That everything is random has to be accepted as a possibility, and the most obvious possibility if honest.

You, and everything you have done and will do, and every thought you have or choice you make, are probably random productions of the infinite.

Every moment you have, and have had, and will have, is

part of an 'eternal' pattern.

CHAPTER TEN

An Infinite TV Screen

The idea that 'everything is information' is key to a modern conception of existence. It has been floating around for a long time, but has become increasingly supported by rational evidence, and is very intuitive for us that live in the digital age. It is often tied in with the idea that 'everything is illusion,' or that we are all living in 'the matrix,' but these ideas kind of miss the point. They imply there is something else, something real to be sought, or an elusive alien behind the program, when illusion is in fact everything.

To say that everything is information is not to say that there isn't a 'physical' world (when correctly conceived and defined), but that everything that exists can be represented in entirety using the system of numbers, and their relations, via the rules of mathematics.

The reason for this is because when looking really close at anything, it is made of units. It is not that there is nothing continuous in existence, but that all continuous aspects of existence are emergent and made of bits, like how waves on the ocean are made of water molecules. Everything that exists can ultimately be created by pixels flashing on and off, the same as a digital image on a monitor screen. Our universe could easily be, as has been pointed out, and is very much the metaphor of the day, a computer.

This computer can be thought of as an infinitely large screen or monitor, like an infinitely large piece of graph paper made of infinite squares, although because it is infinite, there is no need for any software, or even an actual computer.

All you need is for each square to be able to be either of two distinguishable states, such as 'on' or 'off,' or 'black' or 'white,' and you can draw any possible pattern, or will have every possible pattern, generated somewhere randomly.

This includes both colour and multi-dimensional patterns, for even if there are only two dimensional black or white squares, all the information can be there. For example if they are small enough they can depict the different wavelengths of light, and if there are enough of them they can contain all the information which may be displayed in any extra dimension.

This is a pretty simple and an easy to understand metaphor for our existence, which is essentially correct, but it doesn't answer where the screen is, or how there is any movement from one pattern to another, and even more profoundly, who or what is watching. It does not provide a satisfactory conceptual solution to the questions 'Where are we?', 'How is their consciousness?', or 'What are space and time?'.

Although 'space' could naturally be represented by the screen, with every possible moment somewhere in a static distribution of infinite possibility, without 'time' there is no change, nor could there be change, or something which represents our 'experience of time'.

This is where such a simple metaphor fails and needs to be developed.

We could imagine a series of screens, be they finite or infinite in size, one on top of the other, which then make a moving pattern, just the same as a cartoon flip book, or a movie.

Time would be represented by the movement from one image to the other, like a film from a series of photos.

Alternatively we could imagine a focal point moving around a single still image, either programmed somehow, or just randomly from infinite possibility.

The 'movie' metaphor and the 'focal point' metaphor aren't much better than the 'infinite two dimensional black and white television screen' metaphor however, and although in the second example we have made the rather radical move of separating perceiver and perceived, by having a focal point, they are essentially the same metaphor.

There is no explanation as to where the focal point comes from, or what it is, or where the edge of the screen is, as required in a 'movie'.

Although 'space' could be represented by a screen, and time by the ability to move around the screen, or from one screen to another, it doesn't satisfactorily represent where they come from, or how they 'begin,' or what they 'are in'.

CHAPTER ELEVEN

From Foundation to Grid

Instead of imagining space as a photo, and time as playing one photo after another, a better metaphor with which to conceptualise space and time and how they come about or change is to imagine information becoming increasingly resolved. Although there is no information moving, as when looking through a camera lens which is gradually coming into focus, it feels as if something is changing in a forward direction.

That is, our universe, and our existence, is like an image which is being gradually revealed with more and more detail, which is why we experience our sense of movement through space and time.

To consider information becoming better resolved more abstractly, we randomly select any one pixel from the infinite screen, and then imagine this being instantly and randomly divided, a random amount of times. This would result in a grid like pattern, again like graph paper, except the units of the grid can be variably proportioned quadrangles, rather than all perfect squares.

The original pixel which is divided we will call a 'Foundation,' and the resulting grid we call a 'Grid'.

The Foundation and the Grid are linked, because they are the boundaries of our range of information. They are in fact the same information, except the Grid can be thought of as a higher resolution version of the Foundation, which is a single simple unit. Together they form a sequence generator which has only one rule: it is possible one may be divided such that it becomes the other.

For example, and to start with a specifically simple example, the Foundation could be a square, and the Grid could be made of sixty-four squares, like a chessboard. If this were so, there are several different ways the Foundation could divide to become the Grid.

It could be that both sides divide instantly into four, such that together they generate a sequence of only two events, the Foundation as one unit, and the Grid with sixty-four units.

It could be that every side divides in two, at their central point, one event 'after' another, such that there are three events and four 'layers' of information: the original single foundation square, then a pattern of four squares, then a pattern of sixteen squares, and then sixty-four squares in the final pattern. It would be a very uniform predictable and symmetrical distribution or sequence of divisions and resulting quadrangles.

Alternatively it could be that at the first event one axis divides at its middle, and the other axis divides one quarter of the way down its length, such that after the first event there are four rectangles, two smaller and two larger, and after the second event there are a mixture of squares and rectangles, but the same sixty-four squares after the third and final event.

The Foundation and the Grid are exactly the same in both sequences, which share the same possible sequence of four, then sixteen, then sixty-four quadrangles, but they take a different route, which consist of different patterns or shaped quadrangles between the Foundation and the Grid.

Once you have a linked Foundation-Grid pair, there is a fixed 'beginning' and a fixed 'end,' and a fixed 'maximum' and a fixed 'minimum,' within whose range there are only finite possible routes from one to the other, but there are infinite possible Foundation and Grid pairings, each with any number of possible sequences. There are infinite possible Foundations, all of different proportions, and infinite possible Grids.

All the differently proportioned Foundations (which could be a quadrangle of any ratio length and width), coupled with all the different possible Grids (which could be any possible distribution of squares and rectangles), generate a whole range of different possible sequences, between their defining limits.

They could produce a chessboard, but of trillions of units, with its unlikely symmetry; they could produce the Mona Lisa, with its subtle mystery; or they could produce random 'noise,' with very few simple patterns, or, though more unlikely, with no patterns at all.

They could produce the planet Earth, with all its inhabitants, down to every microbe and string of DNA.

They could produce the movement of everything in the universe, as the information changes from layer to layer like screens as it divides, but this would miss the purpose of the metaphor, and be just a hologram which exists in space and time, rather than explaining where they come from.

In this metaphor, the different possible patterns do not exist *in* space and time; they *generate* 'space' and 'time,' with varying measure of success, depending on what sort of pattern they are.

Some patterns generate space and time better than others.

Some have jumpy inconsistent space and time, and yet others have dense predictably consistent space and time.

Those with random noise arguably produce no space and time at all.

In the dividing quadrangle metaphor, how well a sequence of patterns generates space and time is to be correlated to how 'complex' it is.

Complexity is difficult to define, and this definition of complexity needs to be justified, but if valid it means complex sequences generate consistent space and time whilst simple or random sequences do not.

It means, in order to understand the nature of space and time, we have to consider the nature of complexity.

In order to understand complexity, we need to first consider the nature of numbers.

CHAPTER TWELVE

The Most Simple Number

In order to consider the structure of reality, and existence, we do not need to do complex mathematics, or understand complex mathematical equations; we need to consider what numerical distributions are.

Reality has no beginning and no end, and neither do numbers, but finite samples that exist have to start and end somewhere.

Common sense would perhaps stipulate that numbers start at zero and head to the infinite, but actually it is better to conceive them as starting in 'relative simplicity' and heading in the direction of 'increasing complexity'.

Complexity may be difficult to define, but what about 'relative simplicity'?

Information, when expressed materially, is a mark, or a recording, so for a sequence ordered from simplicity to complexity we require 'the most simple mark'.

What is the most simple shape?

A single 'dot' is an obvious contender, which seems perhaps more simple than a quadrangle, given it has one side rather than four, but the circle of its circumference which defines it isn't actually one side, but has lots of sides like a polygon if it exists and can be measured, with infinite sides if it is 'perfect'.

The circumference of a circle with a diameter of one is of course pi, and pi is a number which is so long, as the circle approaches having infinite sides, that it is itself infinite and never able to be fully resolved.

In fact a perfect circle, or pi, is a better symbol for complexity than it is for simplicity, when written to infinite decimal places, and no Grid of quadrangles will ever draw a perfect circle, although the more the Foundation has divided, or the higher its resolution, the more accurately a circle can be depicted, reaching highest possible accuracy at the Grid.

A line is another possible contender for 'most simple shape,' and as it only has one side, and only one dimension, it seems pretty simple. For it to actually exist as a line however, it would have to be infinitely thin, otherwise it would have two dimensions, and be a rectangle, and so like a circle, a line is not so simple after all. For something to exist, and have no infinite measures, requires at least two finite sides, and at least two dimensions. In other words you can't draw a line which has no thickness. Two dimensions are required to make a stamp on existence, because fundamentally for anything to exist, there has to be an x and a y. There has to be something contrasted against something else.

The simplest unit of information is therefore a line with two sides, which is akin to a shape, and more specifically a square, or a rectangle if the sides are not the same size.

The dividing quadrangle metaphor seems suitable therefore as it starts with a single quadrangle, the Foundation, made of two sides, which is its 'most simple state'. Of course each of the sides of the quadrangle must have widths if they are to exist, and can't be infinitely thin, which means they must also themselves have widths, on and on, and so there is an infinite regression right from the start, but this is very fitting and makes sense because our universe is a finite sample of the infinite.

Given infinite possibility, the sides of any Foundation can be considered of random length, as well as not existing anywhere in space or time, so we can imagine them however we want: lying next to each other, or on top of each other, or across each other, intersecting at an angle, and of course as forming a quadrangle.

As well as being represented by the quadrangle they make, the two sides can also be represented by their ratio, with the length of one side compared to the other, which when one side is set to equal one, is what any number actually is.

For example the number sixteen is 'sixteen one's,' which is akin to one divided by or multiplied by sixteen, or a quadrangle with a length sixteen times as long as its width. In this way a number is a shape whose length is measured in terms of its width, and if the length is not proportional to the width, you don't have a whole number.

Whenever we consider anything, there must be a quantity of that thing, such as ten 'fingers,' or two 'electrons,' but what they really are, at least in mathematical terms, is ten 'one's,' or ten 'fractions of one,' or ten 'widths'. With mathematics, there is of course no physical unit of measurement, such as centimetres and it is only the proportion or relative sizes of two dimensions that matter.

In the dividing quadrangle metaphor, the Foundation can be defined by its length, which is the ratio of its longer side to its shorter, and can also be defined by its area, which is its width multiplied by its length.

It makes sense that the simplest of all the possible Foundations, or ratio's, or quadrangles, which is where we are aiming to 'start' with the 'most simple mark,' is therefore the square, where the length, the width, and the area, are all the same and equal to one, which is 'the most simple number,' and not zero.

As the number one divides, which is a square, it divides into two progressions, with the quantity of the number of 'sub-pixels' increasing, heading towards the infinite, and the average size of a sub-pixel decreasing, heading towards zero.

If the divisions are symmetrical, such that they always take place in the middle, and in synchronicity, such that every dimension divides in two, these progressions are inverse versions of each other, with the quantity sequence of new quadrangles being 1, 4, 16, 64..., and the sequence of their relative size being 1, 1/4, 1/16, 1/64...

There are in fact as many numbers between zero and one, as there are between one and infinity, the same but inside out, and as such zero is an inverse infinity which should be considered 'infinity in the opposite direction' because it is also never reached, only approached, closer and closer with smaller and smaller fractions.

Negative numbers should be considered reflections of the positive rather than 'minus' quantities, because in existence there are no such things as negative quantities, only mirror images and opposites, or comparatively smaller positives.

One is the centre, not zero, and it sits in the middle of two infinities.

The number one is everything, and it is the smallest thing.

It is the entire universe, as well as the fabled 'atom,' or that which can be no further divided.

Number 'one' is the most simple number, and is the 'start' of any progression.

Numbers and ratio's and information are about as abstract as we can get, but ultimately the shapes they define are the substance of all interaction, and whether it is considered mathematical or material, or which is the projection and which the substance, is irrelevant. Who wins the long running debate between materialism and idealism makes no difference because they are just different ways of looking at the same thing - an information distribution.

Everything that exists, be it a molecule or a cloud, a quark or a planet, is a collection of 'shapes,' or 'numbers,' and the ratio of their dimensions are akin to the opposing 'forces' which make them.

CHAPTER THIRTEEN

A Dividing Coin Toss

As has been stated, the simplest unit of information is a square, and the Foundation is considered as having two dimensions, each of random length. If they are equal, it is a square, and if they are not, it is a rectangle, with the longer side then referred to as the 'length,' and the shorter as the 'width'. Given that each side is randomly selected from infinite possibility (which is an act that is of course impossible for anything finite, as the finite can't access 'infinite possibility'), you would think the likelihood of them being equal to be infinitely unlikely, but actually the likelihood that our two random dimensions are equal, and make a square rather than a rectangle, depends on how accurately you measure them.

If you could measure to infinite resolution, as if with an infinitely detailed ruler, the ratio would have to be exactly 1.000000… with never ending zeros, which is obviously, when compared to the number of ways they could be different at the same level of detail, very unlikely. The vast majority of ratios in any finite sample would be rectangles.

However, being that this is the most simple unit of information, it has the minimum resolution, which is akin to being measured with a most basic ruler, except of course there is no ruler. The most basic unit is 'measuring itself,' one dimension against the other, and only being one unit of information, it logically only has two options: 'square' or 'rectangle,' with nothing between.

This is because 'information,' be it considered a number or a physical shape, can't just have 'one option,' just as it can't have only 'one dimension'; there has to be a minimum of two options, which together form a single 'unit of comparison'. Without the ability for the information to change between at least two states, there is no contrast, or no opposition, and therefore nothing to say.

You can't really make the statement 'yes,' without the option to say 'no'.

These two options present the ability to be one thing or another thing, like 'up' or 'down,' or 'on' or 'off'. Even if the state is to be always on, it has to be possible to be off for that to mean something, and be 'information'.

Our simplest unit of information, which is the ratio of two random lengths, each of which have no objective size, is formed by comparing one with the other, and determining whether they make a square, or a rectangle.

This is how a quadrangle represents the most simple unit of information, which can be in two possible states.

Given it is the most simple unit, and measuring itself with what is akin to the most basic ruler, half of all quadrangles are deemed a square, and half a rectangle. It would take extra information to skew the odds either way from 50-50, but there is nothing else, and no other information which can be taken into consideration. This means half of the finite samples taken from the infinite would be squares, and half of the samples would be rectangles.

It therefore has to draw the line between what is deemed a 'square,' and what is deemed a 'rectangle,' which is most logically at the half way point between the two, or the ratio 3:2, with all those either side of the margin 'rounded up' or 'rounded down' to its nearest unit.

That is if the width is less than half the size of the length (or the length more than one and half times the width), the quadrangle is deemed to 'be a rectangle,' and if the width is more than half the size of the length, the most basic unit of information is considered to 'be a square'.

At the the most basic resolution there can be no more accuracy than that, and any possible extra detail as to the actual ratio should be considered 'summarised' into a polarised 'square-or-rectangle' distinction.

Every unit of information, at its most basic, is either 'a square,' or 'a rectangle,' and as a finite sample of the infinite, has a fifty-fifty chance of being either.

As with tossing a coin in a vacuum, half of the infinite possibility are 'squares,' and half are 'rectangles'.

Square, or not square.

At its simplest this is the only information an 'infoverse' holds.

CHAPTER FOURTEEN
An Imperfect Balance

So an infoverse starts with a unit of information which registers as a square, for if it registered as a rectangle, it would instead be two squares. This can have a ratio anywhere between 1 and 1.5 (or 0.666... and 1 if placing the smaller side first).

Although a Foundational Pixel which is a square will always 'be a square,' as the Foundation gets chopped into smaller pieces, how accurately it is a square will be increasingly revealed. That is although the exact ratio of the Foundation is fixed and doesn't change, and the level of resolution to which it can be revealed is fixed (because the Foundation-Grid pairing is fixed), when looking at a sequence generated by dividing from Foundation to Grid, because the dividing margin between what is a square and what is a rectangle gets smaller and smaller in actual size relative to the previous division, the measure of the ratio becomes more accurate. The dividing line between what is a square and what is a rectangle remains the same 1.5:1 (3:2), but the quadrangles are getting smaller and smaller relative to the Foundation and so are the margins of division.

For example if an infoverse starts with a 'square,' it could have any ratio between 1 and 1.5, and so for instance have a ratio of exactly 1.244538... with billions more digits (even if they are all zeros). For the first unit of information this would be rounded down to '1,' such that it is 'square,' and each further division can be thought of as adding a decimal place (if it were dividing in ten not two), such that with each division another decimal place is revealed; 1, then 1.2, then 1.24, then 1.244 etc.

In the above example the Foundation Pixel is a rectangle with one side just over 20% longer than than the other, which has been designated a 'square' because it is deemed closer to a square or unity than it is to a rectangle.

This 'square,' with sides of the ratio 1.2, also determines the shape of the unit 'one' in all of its sequences, because whatever the Foundation Ratio is, its area is defined as 'equal to one,' as there is no other measure for the first count.

The 'perfect number one' would of course be a perfect square, up to infinite resolution, or a ratio of 1.00000... to infinite decimal places, whose sides would also equal one.

The likelihood of this 'perfect square' is as mentioned extremely low; infinitely low at infinite resolution, but it can be used as a marker to compare all those which aren't perfect squares, which will be a growing proportion as resolution increases.

The area of the unit is still 'one' however, and the product of its two dimensions by definition therefore also remain equal to one.

This means for the infinite majority of ratio's which are not perfect squares, as the information becomes more resolved, and the ratio is revealed with increasing detail, so each of the dimensions move away from both being one in opposite directions. For our example of a Foundation with a fixed area of 'one' and a ratio which is measured as 1, then 1.2. then 1.24 etc, whilst its lengths could be any 'size,' at their first measure they would be deemed equal, or with a ratio of '1,' and after the second when the Foundational Ratio is revealed as 1.2, they would be be designated as 1.1 and 0.9, and then 1.11 and 0.89 for a ratio of 1.24, and then 1.115 and 0.896 for a ratio of 1.244, which are always the numbers which multiply together to equal one, as accurately as possible within their limits.

The increasing resolution of the Foundation Ratio is therefore also a measure of how far off the lengths of each its sides are from the perfect '1,' such that if one goes up the other goes down.

The Foundational Pixel is therefore not only like a coin being tossed, square or rectangle, but is also like a set of weighing scales. As it divides, the dimensions of the Foundation are measured with greater accuracy, but having no objective size, this is a measurement of one side compared to the other, or how far off centre the fulcrum point is to balance the scales.

This fundamental statistic for any Foundation, its ratio (or how far it is off symmetry), will be most accurately measured at the end of its sequence, or by its Grid, which for a universe, is at the end of its time.

CHAPTER FIFTEEN

Synchronous Functions

In order to more easily consider the patterns that can be generated by dividing a quadrangle, it is useful to consider the possibility of 'rules' or constraints on what is possible. By the nature of the metaphor being used, namely a dividing quadrangle, we have already filtered the possibilities such that there are no additions, multiplications or take aways, and only ever quadrangles dividing.

There is no specific reason for this, for the infinite reality of all possibility obviously has no rules, but every system of information requires them, even if they are believed to be coincidental, and this was our first.

We will now filter the possibilities further and state that all dimensions divide in synchronicity, and only once between each layer. This doesn't mean it isn't of course possible that one dimension divides, and then the other, alternatively instead of simultaneously, or that one side divides in two, and the other in three, or four, or any other number, but we are selecting the most suitable system by which to structure complexity from simplicity. For it is only by beginning with the most simple rules, which combine to make complex programs, that information is organised or ordered into progressively complex systems, and therefore creates a direction.

For the relatively simple pattern of a square dividing in such a way it becomes a chessboard, there are infinite ways this could be done, including the 'fastest' with one jump straight to the chessboard, but the simplest route with a minimum of rules is that every new quadrangle, or 'sub-pixel,' always has both its dimensions divide in two at every layer.

In other words, it is a rule of the metaphor that at every 'event,' every quadrangle divides along both its sides once and only once, such that it becomes four quadrangles at the next layer.

If these also divide not only in synchronicity, but always in the exact middle, this would be where every new quadrangle is of the same proportions as their Foundation, getting more numerous as well as smaller.

The version of this where the Foundation is a perfect square, and all the quadrangles at every level are perfect squares thanks to perfectly symmetrical division, even if there be billions of layers, is as unlikely as any specific sequence, but is also the average of all possibility.

It is right in the middle of all possibility.

It is the grid of perfect squares, at every level.

Because both a Foundation which divides along its own proportions is a useful reference point in order to compare all possible distributions, we can name such a sequence as having 'Proportional Harmony,' and the version in which the Foundation and all subsequent quadrangles are perfect squares as having 'Central Harmony'.

We will take it as a given that all quadrangles divide in 'dimensional-synchronicity' (both sides once at the same time), but not necessarily in symmetry. In that way all the sequences in a particular Foundation-Grid pair are spread around their Foundation Ratio, and all the possible Foundation-Grid pairs, and all the possible sequences, are spread around the perfect square.

Again there is no reason for this; it is just a possibility we are focusing on where all possibilities exist.

If every new sub-pixel has both its sides divide into two pieces at every new layer simultaneously, regardless whether symmetrically or not, the quantity of units or lines marking out each side doubles at each layer, whilst the quantity of quadrangles generated by their intersection quadruples at each layer.

The number of units each side contains as it divides in two again and again is: 1, 2, 4, 8, 16..., where the number and amount it increases by doubles every 'event'.

The number of new quadrangles after each division is: 1, 4, 16, 64, 256..., where the number and amount of increase quadruples every event.

These are exponential sequences, or functions, because every next number in the sequence gets further and further away from the last, so that the rate this gap is increasing at, always increases.

They are sequences which are always accelerating.

These exponential functions are most notably different to linear functions, where the increase of x and y each layer is always equal, the most famous of which is the sequence by which we are most commonly taught how to count, 'plus one' each time (1, 2, 3, 4, 5...) ($y=x$).

Exponential functions are also different to quadratic functions, such as $y=x^2$, (1, 4, 9, 16, 25...), which although increasing by more each time, do so at a steady rate, not an accelerating one (here 3, then 5, then 7... the increase being 2 each time).

The exponential sequence of doubling is represented by the equation $y=2^x$ (and quadrupling is $y=4^x$). This of course lets you calculate the value of either x or y when you have the value of only one of them, and in this case is written such that x is taken to represent 'how far along the sequence you are' (inherently from our perspective in a linear fashion), and to which y responds, although it can be flipped and read the other way round of course, from the perspective of x responding to y, which would be x = square root of y. The variable x in $y=2^x$ is the 'exponent' and when this goes up in a linear fashion, y increases exponentially.

Although the linear 'plus one' sequence seems to us the most simple, the exponential 'always dividing in two' is the simplest of pattern generators, and is akin to a 'beginning,' or the steadiest of back beats, like a cosmic drum, upon which all more complex melodies are built. It is the foundation upon which mathematics is built. It is the Core Algorithm.

Everything can be derived by a sequence whose only program is to randomly divide in two.

CHAPTER SIXTEEN

Multi-Dimensional Gearbox

Time and space are often referred to as dimensions, as are the sides of a shape like length and width or height, or the x and y and z axis on a graph.

Single dimensions are considered lines, as are sequences of numbers, but as mentioned these can't possibly 'exist,' because anything that exists must be finite, and lines of one dimension would be infinitely thin.

In this sense, lines and numbers are actually made of two dimensions, width and length, and there is no such thing as a 'single line,' or a 'single number,' only two dimensional planes.

However the question then arises as to whether these quadrangles have any height, and whether information must actually be in more than two dimensions. According to orthodox theory, and common sense, there are three dimensions of space, and one dimension of time (although 'string theorists' for instance believe in ten or eleven dimensions). Our consciousness obviously sees a three dimensional space, and a one dimensional time, such that we can't imagine any sort of other space we could move through, or a time in which we could move left or right, so how many dimensions are there? This of course depends how you define 'dimensions'.

A dimension is a unit, or collection of units, of information. In this sense it is also a function, as in an equation, or a sequence of that equation, and every number in that sequence. It is also a line, or a shape.

It is basically any form or distribution of information.

Mathematically, there can be any number of dimensions, as there can be any number of numbers, from zero to infinity, and even all the fractions between (which sounds rather strange but if one dimension is half the size of another, which is defined as 'one dimension,' in that way it makes sense).

This seems to challenge the idea of three dimensions of space, but actually it doesn't because there are not 'three' dimensions quantitatively, but three types of dimension.

These three types of dimension, for space, are often represented by a line said to be in 'one dimension,' a square said to be in 'two dimensions,' and a cube said to be in 'three dimensions'. The reason for this is obvious, and makes sense of course because this is how us humans see the world, but it can also be said that all these different 'dimensional types' are in fact 'two dimensional,' because all information is 'two dimensional'.

This might seem a tad abstract, as if our existence could be printed on a piece of paper (or be a three dimensional hologram of the two dimensional edge of space as some scientists believe), until you realise one perspective does not exclude the other. The information that makes us could be two dimensional, but translating the functions into three 'physical dimensions' a necessary aspect of our existence.

Indeed, because this possibility is integral to our perception, it can be considered that living beings are what 'translate' the two-dimensional information into more complex forms, with our brains generating a three dimensional space from a linear perspective, just as we do when watching a film, but much deeper because there is no need for 'translation' as our brains and the 'three dimensional' world, like the chicken and the egg, came together.

Thinking about multiple 'dimensions' does become a bit of a semantic minefield however, as well as a conceptual one of course, because we have to remember one and two or three dimensional patterns of information are all different types of 'two-dimensional' information, no matter how many dimensions they have! In order to try and untangle this knot, from now on each dimension of information can be described as having two 'axis,' even though these axis are of course dimensions, each with their own axis, as any sequence or unit of information is, entangled together in the infinite.

In this way, when a unit of information (which like all information has two 'axis'), generates a sequence, patterns of possibility can appear as 'squeezed into one dimension,' or 'spreading out in two dimensions,' or as 'expanding into more dimensions,' which we interpret as a 'third dimension'.

These three different types of dimensions, as far as our 'space' is concerned, can emerge from our very simple sequence generator, a dividing quadrangle, which remains in two dimensions, or as one dimension with two axis.

Information which appears to be in 'one dimension' is information which is sequenced such that it expands uniformly along either of the axis, vertical or horizontal, in a straight perpendicular line. It is as if the information is dividing on one axis and not the other.

Information which appears to 'spread out in two-dimensions' is information which is sequenced such that it can expand along both x and y axis in proportional synchronicity as increasing area. That is the axis x and y always divide at the same place and same time creating a single dimension of uniform expansion.

Information which appears to be 'three-dimensional' and have 'volume' is information which is sequenced such that it can divide on both axis but at different rates, or 'independently'. That is the x and y axis can divide at different 'times' or different 'lengths' and so appear to expand into multiple dimensions, consistently creating a 'third dimension' when so aligned.

All these three different types of dimension are 'two dimensional,' because ultimately all information is 'two dimensional,' but they are three different ways of sequencing it.

They are of course the three different types of function; linear, quadratic, and exponential respectively. That is linear is 'one dimensional,' quadratic is 'two dimensional,' and exponential is 'multi dimensional' or 'three dimensional'.

Rather than utilising the three different functions as lines, squares, and cubes, the different methods of sampling information can also be compared to changing gears, with a processor, like an engine and a gearbox, able to shift up and down levels. Except these gears are functions, and unlike gears in a car, we as complex beings are able to use all gears to process information simultaneously, whilst altering the proportions in each, rather than having to move all 'energy' from one to the other.

Bringing them altogether into one complex sequence with a steady 'forward motion', whether this be generated randomly or the consequence of a complex perception, is where the fourth type of dimension comes in, 'time'. It is not just another function, but more like a clutch, and able to align the different gears.

CHAPTER SEVENTEEN

Axial Resonance

The previous chapter attempts to describe the idea that there is a lattice of interconnecting 'dual axis' information, made of up to infinite 'dimensions,' which can blend or divide and expand in three ways, often thought of themselves as 'dimensions'.

Each of these dimensions, even if lines and 'one dimensional,' must also have widths, which must also have widths, on and on into the infinite, but as already mentioned this is apt because we are taking a slice of the infinite. The sides of the Foundation not only have widths, but are themselves the widths of much 'bigger' dimensions, such that the Foundation is a single finite link in infinite possibility.

When we think of the side of the Foundation square, which has been designated as having a length of 'one,' it is a line with a length much greater than its width, and which must have a length which determines the value which this 'one' represents.

In order to measure this 'length,' we count the number of units the Foundation axis has been divided into. As the Foundation it is assumed to have a length of 'one unit,' but the sequence generates the possibility of 'two units' and 'four units' etc, doubling at each division and maximising with the Grid, even though it isn't getting longer in a physical sense, only an informational one.

In this way, the maximum possible length information wise is determined by the size of a Grid unit, or how many times the side of the Foundation can be divided by a unit of Grid size.

As for the size of the width of this length, which must have a value as nothing can actually be one dimensional or infinitely thin and exist, we can assume the line is as thin as is possible to measure within the sequence, which is of course also determined by the length or width of one of the Grid squares. The length of Foundation axis can therefore also be considered at its 'maximum' by dividing its length by its width, or the ratio of its two sides (the essence of what a number is).

We can also consider the line that is doing the dividing must also have a length and a width, if it is to exist, and we can assume by definition it is also as thin (and short) as can be in the sequence and determined by the size of the Grid units. This makes sense because it is also the case that at the Grid, the thickness of the line dividing is equal to the length of the line being divided, such that the sequence can divide no more.

The accuracy of the measure of the Foundation length, or how many times it has been divided by a Grid unit or its own width, is dependent therefore not only on the consistency of Grid units all being exactly the same size and with Proportional Harmony (minimising any inaccuracy without rounding up or down by dividing into whole numbers not fractions), maximising the maximum if they have Central Harmony and are all perfect squares, but also that the line dividing has a consistent width equal to the size of the Grid units, as well as the width of the line that makes the

Foundation's side.

These possible alignments act like a layer of information which determines how well the numbers in the sequence are 'focused'. We can call it Axial Resonance, and it could for instance be a way to understand how human's can sometimes feel 'in the zone,' whether that be when playing a sport, or creating some form of art. All the units of information generating the experience, the divisions and the divided, are for a time lining up and harmonising in proportionality.

CHAPTER EIGHTEEN

The Apex of Complexity

The different possible Foundation to Grid relationships, each with all their different possible sequences and symmetry, can themselves all be sequenced, or ordered, such that they are organised and can be assigned numbers, like books in a library.

Books tend to be organised by genre, and author alphabetically, which is relatively easy because although genre may sometimes require a subjective summary, every author has a name, and every book has a title.

For all the different possible patterns generated by dividing a quadrangle into lots of little quadrangles, we can order them by their complexity.

Unfortunately, although it is easy to intuitively recognise the difference between an obviously simple system and an obviously complicated one, complexity itself is difficult to succinctly quantify. It is not like you can put a pattern on a set of scales and take a reading of how much complexity it has. We can distinguish two main ways information can become more complex however. It can become more numerous, and it can become more accurate. These have often been referred to as the variables of 'quantity' and 'quality,' and both are wanted for 'meaningful information maximisation'.

The quantity and quality duality is an extension of the fundamental duality of the most simplified information, which must have two states. As mentioned, a number is the ratio of a rectangle, length over width. In a world of pure numbers, the 'quantity' is the length, and the 'quality' is the width. The width is the 'measure,' and the length is the quantity of the measure. Although with one unit of information it is 'this' or 'that' ('on' or 'off,' or '1' and '0'), with a multitude of information it becomes 'so many of these,' and 'so many of those,' where 'these' and 'those' are 'things;' as in 'three' 'apples'. The 'quantity' of the thing is three, and the 'quality' of the thing is that it is an apple. If the information in 'binary' can be '1' or '0,' then you get 'three '1's,' or 'three '0's,' where the state of '1' or '0' is the 'quality' of the information.

In the dividing quadrangle metaphor, quantity is equivalent to the number of pixels on a screen increasing, and quality can be thought of as how 'aligned' the dimensions are, or proportional to their width.

If we consider both the quantity and quality of information, or it's general 'accuracy,' it would make sense that the more complex a sequence is, the more accurate it is, with both higher quantity and quality. If we had a 'complexity scale,' the most simple pattern would be the least accurate, with a low complexity rating, and the most complex would be the most accurate, with the highest complexity rating, on and on towards infinite complexity.

It is a simple correlation, as with most scales, where a larger number represents increased complexity, despite the fact we haven't accurately defined or quantified complexity.

However, and just focusing on the quantity aspect of complexity, although having more units does correlate with the possibility of greater complexity, such that it does make some intuitive sense that 'two is more complex than one,' or 'two million more complex than three,' it isn't necessarily the case that two is twice as complex as one, or ten twice as complex as five etc. In this sense, the 'expansion of complexity,' or the rate at which a pattern gets more complex, could increase, and thus be better represented by a quadratic or exponential scale, especially when also considering quality.

Although we can't define complexity, and therefore determine which line or function or 'type of dimension' should best represent a maximisation of complexity, we can say that when the next number in any sequence maximises complexity, the sequence is taking the 'apex of its possible complexity,' as a car may take the fastest and most energy efficient apex round a corner, or what we may call 'The Apex of Complexity'. Indeed, complexity is presumably not programmable, and therefore impossible to define be it with words or a singular mathematical equation. It is however still a 'pathway' that 'exists,' which must utilise all types of function, and encompass the full spectrum of possibility. Every moment it is further beyond our reach, yet like the sun or the moon providing direction.

The pinnacle of complex possibility, at the end of a limitless apex, is a singularity far beyond even our imagination, yet it structures any sample of randomness into a 'numerical sequence'. Like magnetic poles influencing iron filings, or even the idea of divinity to the devout, the progression from simplicity to complexity is the quintessential organiser of all random possibility.

CHAPTER NINETEEN

A Perfect Average

So a very simple coin toss centres all reality, and generates all possibility.

With our dividing quadrangle, as we have two dividing dimensions (or a single dimension with two axis if you prefer), we essentially have two tossing coins, with their waves of possibility overlapping and interacting, or generating the possibility of appearing to do so. These are like waves on a pond, with pebbles being dropped in two locations, except they are statistical waves and any interaction they have, as well as any patterns they create, such as spreading out from a central point, are completely random and coincidental.

The sequence of information displayed is determined by all the different possible ways a quadrangle can divide from a Foundation to a Grid, and represented by the sequence of 'information layers' or 'sub-grids' (made of 'sub-pixels') and the pattern of different squares and rectangles they produce. With this, and with enough detail, our whole world and universe could be encoded.

In order for the data to be 'played' however, or for their to be a 'machine which can play it,' such as our brains, there needs to be space and time. There needs to be somewhere for the 'information levels' to exist, where the numbers of a sequence are enacted. There needs to be physical substance.

How we move from hypothetical information reality to a physical existence is the crux of any 'infoverse' dilemma. How on Earth does 'possible information,' which exists nowhere, end up becoming a physical world, with living beings who think, or for example experience taste, as we clearly do?

Objectively, it makes most sense that all the data in every existence is timeless. All the possible sequences 'exist as possibility' forever, 'outside' of any existence, and objectively there is no 'direction'. Nothing begins, and nothing 'moves'.

Subjectively however, there are lots of directions, and movements, at every possible angle. There are currents, or prevailing winds, and tensions between opposites, and it is these which create the 'flow of time' - which we think of as moving us all in one direction; and the 'region of space' - which we think of us moving through in many directions.

These currents seem to have an origin. If we rewind time, everything goes back to a pin head, to the Big Bang, or the beginning of space and time, which came from a giant cosmic explosion in what was apparently nothingness.

This seems to be the beginning of time and space, even though given the density of space at the time, squashed as it was into a dot, time itself would have been barely moving, at least compared to our time now. What would have been much much faster than the blink of an eye, compared to today, would have in fact taken the age of the universe, compared to today. The closer you get back to the first moment, the larger it becomes, as if approaching light speed, or accelerating up an exponential sequence, because the Big Bang is continuous, like a constant furnace at the centre of the universe. It is better to imagine it as always happening, or happening again and again, with it being only our perspective that gets a little further, every moment, as we surf its layered dimensions.

Space and time as we perceive them do begin, and expand, and move forward, this is obvious to everyone, but from an information point of view, they are both the consequence of the way numbers may be sequenced.

Physics and chemistry wise all energy is derived from an inbuilt response to balance opposition, which ironically results in opposing forces.

There is magnetism, with its north and south poles, and electricity, with its plus and minus, which were of course discovered to be related and together are now known as electro-magnetism.

There are two nuclear forces, the particles of which have positive and negative charge, repelling or attracting each other allowing only forms of matter and mass which are balanced and stable enough to exist in space and time, more or less briefly.

And there is gravity, which seems to fight energy itself, wanting to retract all that exists back into a tiny very heavy dot, the conflict between the two typified in our universe by a star, which may go one way and explode to Supernova, or the other and retract to a Black Hole.

For anyone who believes existence to be information, these four forces are themselves statistical patterns.

They can in fact be considered as material representations of the four forms of mathematical operation, namely division, addition, subtraction and multiplication, and they balance the same way an equation balances around the = sign.

It is also the essence of a random distribution that it balances around an average.

Toss a coin again and again and the odds are you will approach a more and more exact even split of results, with a greater sample generating greater accuracy.

All statistical direction is towards the average.

'Approaching an average' is 'seeking balance,' which is the essence of energy and also gravity.

All the mathematical interactions, or physical forces, seek balance.

The law of averages is the main program. In some ways it is also the pivot at the centre, or the equal sign, or the number one.

Nothing can escape the law of averages.

The average for any universe is the ratio of its Foundational Unit. It is its Foundational Ratio. This is its central resonance, which is very unlikely to be perfectly average.

A rather profound projection of this is displayed by taking a random unit of rope, and asking lots of human beings to guess its length; it is highly unlikely anyone will be exactly right, but the average of all the guesses should be more and more accurate the more people asked, like tossing coins.

Another is the tendency for polarisation of opinion, often necessary in order to simplify information and make a decision, except perhaps there is no decision, and no individualism, with us all a part of a connected distribution.

In this way we have left and right in political opinion in the same way magnets have a north and south pole.

In this way, there are different genomes and personalities and even decisions spread around an average, just as there will be different forms of anything, like planets, or rocks, or cattle.

In this way, we are all statistical patterns, spread around an average, approaching that average by drawing it more and more into focus, the perfect version of which is unlikely to actually exist.

'Approaching the average' may sound underwhelming,

but in some context it is actually the same as 'perfection'.

Somewhere in the spread of all possibility is a perfect - or an average - version of ourselves, as well as our species, our planet, our society, our nation or any other aspect of existence able to be defined or perceived.

We are all aberrations of this perfection, just the same as light may be refracted around the centre of a lens.

CHAPTER TWENTY

Resolving a Zoom

With the dividing quadrangle metaphor, patterns can develop in any sub-grid from layer to layer, be it a chessboard or a Mona Lisa, where they can remain, change, or disappear. Patterns which survive from one layer of information to the next, as the quadrangles divide again and again, maintain linear consistency even though they are built on exponential patterns, which gives the impression they are moving, although for now it should be remembered this movement is not in space or time, but hypothetical possibility. Whilst it is tempting again to equate this movement with a movement in space or time, this is us projecting the need of our conscious focal point to move onto what are still so far 'static possibility statistics'.

Patterns which could from some hypothetic external perspective 'move down the layers' will either be 'expanding' on the x and y dimensions as the division increase the number of pixels, or 'zooming' into x and y by remaining consistent in their pixel quantity.

It is like an image on a screen quadrupling its resolution in jumps, such that you can either improve the overall image quality, or digitally zoom by maintaining the same resolution.

Either a pattern maintains itself as the quadrangle divides by becoming made of more sub-pixels with each new layer, or we can chose a pathway down the layers by focusing on a single quadrangle at each layer. In the latter case whenever a quadrangle divides in four, one of those is randomly selected, and when that divides in four, one of those is randomly selected, and so on, all the way down to a Grid pixel at the final level. Instead of becoming increasingly resolved, it would be heading in the opposite 'direction,' maintaining the same resolution, but becoming four times smaller in terms of relative area at each level. In some sense, instead of utilising the x and y dimensions to 'expand,' it would be utilising them to 'zoom'.

We therefore have two directions of travel for a pattern.

They can stay where they are and become more resolved, or they can zoom with the levels and stay at the same resolution.

At each level of information it is strictly one or the other: 'Resolve' or 'Zoom'.

Both are increasing complexity, because just by surviving from one layer to the next a pattern is beating the odds, but they are doing so in different ways.

The speed of resolution is quadrupling.

That is the rate at which the number of units increases with every new layer.

The speed of zoom is quartering.

This is the average relative reduction in size for a sub-pixel from one layer to the next.

CHAPTER TWENTY-ONE

Framing A Focal Point

If a pattern is perfectly consistent from one layer to the next, and utilises every new sub-pixel to remain so, it is maintaining itself at a greater and greater level of accuracy with each new layer requiring more and more sub-pixels to be correctly and specifically aligned.

In order for a pattern to maximise its resolution this way, or the number of sub-pixels it has, it would have to incorporate as many sub-pixels as possible, such that the Proportional Harmony pattern, which divides symmetrically at the same proportions as its Foundation from the very first layer, would be this pattern.

Any deviation from the ratio of the Foundation would result in the pattern changing to something new, and thus discard the information that had 'been,' reducing the length of the consistent sequence.

In order for a pattern to maximise it's zoom, it would also have to maintain its pattern from the very first layer, such that every new pixel which was selected at every new layer had the same ratio that it did.

Again, if the new pixel 'chosen' at each new layer is of exactly the same proportions of the first, to an increasingly accurate measure, its 'length of zoom' will be maximised, whereas if it isn't, there will be change to the pattern and something new will emerge which then may or may not be consistent.

When a pattern is zooming down the layers and maintaining the same resolution compared to a pattern which is becoming more resolved, it has to 'chose' which of the four new quadrangles to 'focus' on, rather than just have them 'randomly arrive'. This selection from possibility can of course just like the divisions be random, but the zooming inherently creates a 'Focal Point'.

Patterns which become more resolved, don't so readily create a single Focal Point, being that they are always made of an increasing number of quadrangles, but they can be thought to have a 'Focal Range'.

The Focal Point of a Focal Range is an average of all the pixels, which becomes an increasingly accurate estimate of the Foundation ratio.

If a pattern has had some mixture of Zoom and Resolve, such that it has not always selected one quadrangle at each layer, nor remained the entire Foundation, any of the new Focal Ranges can be referred to as a 'Frame'.

A Frame can zoom and then resolve, generating complexity by 'selecting' possibilities, but it sacrifices 'duration' in the process.

CHAPTER TWENTY-TWO

Diverse Harmony

If a pattern is to maximise either its resolution or its zoom, it must maintain its proportions, and be harmonious, but neither possibility is likely to result in a maximisation of complexity, which would surely require a pattern which mixed zoom and resolve.

Another obvious way to increase complexity is to generate 'collections of patterns,' with each layer full of various quadrangles that differ in ratio to the Foundation, which could then consistently resolve or zoom. Diverse collections of patterns could also loop sequences down layers, repeating not just every layer, but every two or three or any other number, such that they spread along the information layers not just across x and y. In these cases the very act of change, when repeated, becomes part of the pattern, as for example if a quadrangle gradually oscillated from an extremely thin rectangle leaning left to an extremely thin rectangle leaning right, via a square in the middle, and back again.

All the different possible patterns or collections of patterns can be incorporated into the definition of a 'Frame,' such that there is diversity from resolving and zooming, spreading out in x and y and down the layers, as well as diversifying proportions of the increasing number of quadrangles.

This mixing resolution and zoom, with a sprinkling of random and repeating variations, brings to mind the structure of music, with its varieties of tempo and pitch, which is not surprising given the well known connection between music and mathematics.

Not only are musical notes ratio's of amplitude to wavelength just like the patterns of the quadrangles which as a multitude can create waves, but a zooming pattern with few quadrangles can't reach the same levels of diversity as a resolving one with many, a bit like how an orchestra can create a greater variety of possible sound compared to a soloist.

Additionally, if a pattern is spread over more than one level, the more it spreads, and the longer the sequence which loops, the less number of times it will be able to repeat before the end, much like how a long verse can't be repeated as many times as a shorter rhythm for a fixed length musical piece.

In the case of our dividing patterns, there are however some obvious differences. A diversely resolved pattern for instance, unlike a piece of music with lots of different pitch notes, can't maintain its diversity right from the beginning and until the end, but instead has to build up and peak in the middle before gradually harmonising approaching the finish. It can't be diverse from the 'start,' because it takes 'space and time' to build up diversity, or possibility, whereas a piece of music can start with everything and end with everything. Also, with information in the dividing quadrangle, the greater the 'complexity peak,' the shorter the duration, which is as if with all that extra effort, some instruments would be worn out and unable to play further, whereas a more harmonious piece of music would be able to carry on longer. With the quadrangle metaphor, as with the universe and existence, there is a trade off between complexity and duration.

Unable to exactly define complexity, we could argue that complexity is better represented by harmony than by diversity, or better generated by resolving rather than zooming.

Most likely of course, as with quality and quantity, complexity requires a mixture of harmony and diversity, as well as resolving and zooming, in order to be maximised.

CHAPTER TWENTY-THREE

Energy Tree

Every metaphor is a simplification of reality, and as simplicity can never detail complexity, so there will never be a single metaphor which can detail all of reality, or an equation that can program existence to infinite accuracy. It is this which will always separate the human born from reality, and the machine born from humanity.

In that vein, there are many metaphors used to describe 'energy,' including the metaphor of mathematics, each adding their own view, and it can be helpful to sometimes think of the original Foundation Pixel not as a quadrangle dividing, but a bucket of water molecules being poured down a branching pipe.

If we go back to the example of a proportionally harmonious universe, which is maximising its increasing resolution by remaining symmetrical from Foundation to Grid, then all the quadrangles can be thought of as running water dividing like a river delta at the ocean. The water flows down first one pipe, which then splits into four offshoots, and then sixteen, and then sixty four, etc., or into four more at every node. There is always the same amount of water, but as it divides so the likely quantity in each pipe reduces.

The likely reducing quantity in each pipe is like reducing water pressure, which can be represented either by the length of the pipes getting smaller, or the angles they are at getting shallower.

With a harmonious universe, or the average of infinite possible sequences, all new pipes after a node would run at the same angles, getting shallower at each split by the same amount, one quarter of the previous angle (or having their lengths decrease at each split, one quarter length), and when the water runs out, there would be exactly one molecule in each pipe, each of the same length or angle.

With a diverse universe, the branches would divide at different angles, or have different lengths, always getting shorter or shallower, but by different amounts, and at the end they would either all still have multiple molecules, or some would have one and others multiple, or some branches would be shorter than others which carried on dividing many nodes further.

Using this metaphor we can say the information available at the beginning is that there is one bucket of water, but there is no way of measuring how many molecules there are within it.

There is one unit of 'potential energy'.

After one division, or level of information, it is known there are at least four units, but it is not known what each unit is worth. There are four units of 'kinetic energy,' which can express themselves as a square or a rectangle, or a light which is on or off, but it is not known how much potential energy is left in each.

By the time the process ends, it is known exactly how many units there are, as the maximum amount of kinetic energy has been expressed, and the value of each is at its most accurate it will ever be, with a value of 'one Grid Unit,' give or take only half a Grid Unit, together the most accurate measure for how many molecules were in the bucket.

Of course, our bucket of quadrangles isn't actually made of water molecules, with pressure, it is made of units of data, which are able to display information, the possible sequences of which can be organised by probability, or the unlikelihood of complexity.

What is more, and closer to a metaphor of electricity with switches or electric gates perhaps than water and pipes, but of course not perfect with either, each of our units of data have the possibility of two opposite states.

At each of the gates where the water or current splits, each 'data unit' divides into four different units, and each of these can either be classified a 'square' or a 'rectangle,' a bit like positive and negative.

The data units, or water molecules, can therefore also be compared to spinning coins. Each time one goes through a node, or divides, it spins and splits into four new coins. These can be either heads or tails (a square or a rectangle), but they are rather special coins, in that they have as a group a shared symmetry. When dividing there can be four 'heads,' (squares), or four 'tails,' (rectangles), or two heads and two tails, but there can't be three heads and one tail, or vice versa. This is because they are actually quadrangles, or paired dimensions, and when they divide they divide into two pairs which must have matching ratio's even if they are of different size.

What is more, unlike a coin which has fresh odds with every spin, these accumulate odds, such that if you have had a tail, you are more likely to get a tail next, and if you have had a head, you are more likely to get a head next, because a thin rectangle dividing in four is more likely to make four new rectangles than a square dividing in four which more likely to make four new squares. It is akin to a statistical form of momentum.

There is also a general drift in the direction of the ratio of the Foundation, or the 'harmonising of the quadrangles,' which is why the Foundation Ratio becomes more accurately measured, for if the sequence makes it to the end, all coins would be the same as the Foundation, as determined by its highest level of accuracy.

CHAPTER TWENTY-FOUR

The Central Perception

It is possible to comprehend a sequence of numbers or dividing quadrangles which represent a pattern or structure of information, and it is also possible to comprehend that the shape of this information generates multiple possible pathways, any one of which (or many integrating as a Frame), can be pursued by an appropriately varying Focal Point, which could from an unexplained vantage point and despite being coincidence appear as flowing energy. Yet it remains very difficult to fathom how dividing quadrangles can explain what our own perception is, which still seems a huge leap in conception.

One big challenge for any attempt to comprehend how space and time 'came about,' and with them consciousness, the nemesis of all metaphysics, is that being a pattern of space and time, we can't imagine anything outside of space and time. Given that it is the integration of two dimensions, x and y, which are being considered as the means to 'generate' space and time, or 'enable them to exist,' this is a bit of a problem. The dimensions can't exist without space and time, or we can't imagine them 'existing' without space and time, and space and time can't exist without the dimensions.

A key question arises:

"Are space and time generated by the sequence of 'energy moving,' or is a sequence of 'energy moving' impossible unless there is already space and time?"

This is of course like the chicken and the egg situation, with the answer obviously that they come together, because they are the same thing.

With a computer program, electricity is needed to make it work.

With existence, the structure of possibility generates the 'energy' flowing down it.

All energy in our universe, whether it be the smallest unit or largest collection, can be thought of as Focal Points, or a Focal Range, and we are of course a Focal Point, or a collection of many resolving and zooming Focal Points, as conscious thoughts, and physical individuals, as well as parts of the greater pattern that is our shared universe. All of these zillions of Focal Points, or 'energy points,' have different pathways, but have coalesced in such a way that we co-exist.

It is the arrangement of information which generates space and time, as much as it is space and time which rearrange information, and in order to imagine what makes the program of existence 'run,' and the code written by random possibility capable of being executed, as it clearly is as testified by our existence, we have to imagine a zooming Focal Point and a resolving Focal Range being brought 'into focus,' as with a telescopic lens.

The attempt to distinguish between a pattern and the energy which runs it, or a perception and any chemical or instinctive response, is when put into our own context as living conscious beings, the long considered philosophical question of solipsism. That is, whether perhaps our mind is our universe, and we are at the centre of it, 'constructing it'. Perhaps it is each of us that runs our own universe; that a brain such as ours is required in order to perceive space and time as we do.

Everything comes together, our mind and our universe.

In a very real sense, we are each our own universes, but our universes overlap, and so we also have a shared universe.

Our pattern is a complex pattern of many integrated patterns.

A milder version of 'Am I the centre of the universe?' (which is in essence the mystic's question 'Am I God?' or 'Is there only one being in existence and it is all of us?'), is the question as to whether there is any 'time' or 'space' without consciousness.

When we wonder if there is a universe without us, or without a consciousness experiencing it, or without any life, we are wondering if there is any space and time or 'mass' without mind, or without life. Our perception is integral to our view of our universe, and whilst an insect has a very different view to a human, they will still perceive 'space and time,' but this can not be said for a rock, like the Moon.

It is the varying focus which enables energy to exist, as much as the flow of energy which enables perception. The flow of energy, as we see it, would therefore not exist without our perception.

Life is not different from chemistry in any essential way, but the perception of a living being, and especially a conscious being, is obviously different to that of non-sentient or non-living mass. Our increased jump in complexity, which is what life is, was a paradigm shift, or a cosmic shift, in terms of the generation of space and time.

The Moon, or any mass which is not living, exists in spacetime, but does not experience 'space' and 'time' as separate dimensions like we do.

The universe as we see it does not exist without us, and it is therefore a complete fallacy to imagine an empty universe like ours without life, as depicted in all astrophysics documentaries, because without life a universe such as ours can't exist. There is no space or time without life (or at least not space and time as we experience them). There was no 'Big Bang' without the human mind, and each of us had our own.

To state that our minds generate space and time sounds like mysticism but is perfectly valid. In fact science which has separated them, and cast us off from our universe, has in this way deluded humanity. Each of us sit at the centre of our own universe, overlapping with others who sit at the centre of their own universe's, and when we die, our universe will end, yet what it 'has been' will always be.

CHAPTER TWENTY-FIVE

Spacetime is not Space and Time

It is very difficult for us to truly understand how space and time came about, because we have a refracted view of existence, being a complex perception. We need to simplify space and time, in order to understand them, but we cannot just simplify ourselves, as you can't just simplify a machine, you have to reverse engineer it. We have to unpick the complexity, to fathom how it developed.

In order to understand how space and time and mass all relate, we shouldn't begin by studying them separately, which is the complex part we naturally do. In order to simplify them, we have to put them back together.

'Space' is for most of us the physical universe. It can be more or less dense, and where it is less dense, it is more spacious, and where it is more dense, it is less spacious, and has more mass. Mass and space seem to be the same, or overlap, but when we look at our universe it is obvious to see the difference between them. They are graduated from the very dense and massive Black Holes to the very 'deepest space,' where there is some mass, or some gravitational field, though it is very weak. They are also separated by our beginning in the Big Bang, which we think of as a very small dense and massive space exploding, passing through our middle stage where we are with varying space and mass, heading to our end which is thought to lie on a tightrope between a Giant Whimper and a Big Crunch, the first being where we carry on expanding getting thinner and thinner, with all mass reaching for nothing; the second where after a maximum stretch, we contract again with increasing density back to another small squashed dot and possible Big Bang.

Space can have more or less mass, and so we naturally think perhaps mass is an ingredient of space, where some space can have more mass than other space. This is however wrong. If we think of dense mass taking up less space than spacious mass, we are also wrong. Einstein explained how dense mass is contracted space, not contracted mass within space. Mass contracts space itself, it does not contract within space. There is relatively more space, in dense space, than in spacious space!

It is difficult to grasp.

Rather than space having more or less mass, it is more like mass can have more or less space. It is more like space is an ingredient of mass.

But this is also wrong.

Time on the other hand is the inevitable forward motion which we all experience. The rate at which we experience time can vary with metabolism, so a fly experiences time slower than a human, but time, including metabolism, is also related to mass, like space, because it flows at different rates depending how much mass is present. The shape of space, and how contracted or spacious it is by mass, is correlated to the rate of time. Where there are dense planets, or stars, or Black Holes, the slower time travels. If we could see time, it would follow the same distribution of the various densities of space, or mass.

You may think like with space, that mass is an ingredient

of time, or that time is an ingredient of mass, but again both are wrong.

Time is not an ingredient of mass, any more than space is an ingredient of mass, but both space and time do come from mass.

Mass seems to effect both space and time, if in inverse directions, because it contracts space and stretches time, so we could consider that the presence of mass separates space and time, by contracting one, and stretching the other.

However this also isn't the case.

Mass *is* space and time, except it is them 'before' they became separated into space and time.

Together, as mass, they are often referred to as 'spacetime,' but this is a very misleading term, because it implies spacetime is made of space and time, when there is no space nor time in spacetime.

To best consider what happened before there was space and time, which is of course an oxymoron and 'attempt to imagine existence without space and time separated' may be better, is to imagine a universe which has only spacetime, which we can just call mass.

If mass was the same everywhere, right across space, dense planets, and stars, and Black Holes, would become much bigger, and deep space would become smaller. Space would in fact be the same everywhere. If we could see the rate of time, with even mass everywhere, where there are dense planets, and stars, and Black Holes, it would speed up, and where it is now faster in deep space it would go slower, such that it would even out and time would go at the same rate everywhere.

If we levelled mass out, space and time would have the same rate everywhere, but there would be no differentiation, or separation of things which move.

Only if mass has variable density, can it generate the

possibility of a process which differentiates it into space and time.

The complex process of separating variable mass into space and time as we experience them, which is also an inherent part of their being variable mass, is of course 'being alive'.

CHAPTER TWENTY-SIX

A Gravitational Grid

The presence of mass is explained in modern physics by a gravitational field. Where it is dense and its lines are close together, there is a lot of gravity, because spacetime is warped. Mass is said to warp spacetime.

Much simpler is to forget space and time for now.

There is just mass, and it comes in various densities.

Utilising the dividing quadrangle metaphor, the gravitational field is the dividing grid, and the number of quadrangles in a Frame determines how much mass it has.

Where the lines are closer together, and the quadrangle is smaller, the mass is denser. Where they are further apart, it is less dense.

With harmonious patterns, the grid maintains the same proportions, so whilst the quadrangles get smaller, and the lines get closer, they remain equidistant from each other, with a constant spread of mass, even though from the point of view of a Frame which is not zooming but resolving, it is getting more and more massive.

With diverse patterns, some lines are closer than others, which creates a mix of different size quadrangles, and different regions, with those of smaller quadrangles in a

Frame equating to relatively dense mass.

Because the intersection of x and y create quadrangles, the number of which within a Frame equate to its mass, it can be tempting to think because x times y is mass, x could be space and y be time, and together they make 'spacetime'.

But this is again where we tend to put the cart before the horse.

The dimensions x and y are not space and time. The dimensions x and y are just dimensions of possible data. Their intersection can only be thought of as making mass (or kinetic energy, or information, or spacetime).

It is the process of separating the product of x times y ('xy') into two new dimensions, which are completely different to x and y, and may as well be labelled 's' and 't,' that creates 'space' and 'time'.

These two new dimensions can also be displayed as perpendicular to each other, just like x and y, so the confusion is even more understandable, but they represent a very different relationship within the increasingly complex sequence that leads to space and time. The separation, or differentiation of mass, into space and time, is not the reverse process as making mass from x and y.

In an increasingly complex pattern, mass, or kinetic energy is always increasing, but as mentioned it takes another a whole new level of complexity to separate mass into space and time.

Our universe actually began with our birth, and our shared universe, or overlapping universes, began with the first form of life.

Whenever we think of the universe without life, metaphorically speaking, we should not think of mass as moving in space and time, but simply of mass as increasing and diversifying, from a small seed into many patterns. We can't imagine this without utilising our experience of space and time, and of course there can be no diversity at all without space and time, but if there could be a simple or 'chunky' form of space and time, there would be an explosion of mass into a variety of densities, some getting denser faster than others, but everything overall getting more massive, as more and more of the potential energy becomes kinetic, and there are more and more quadrangles, or an ever more refined and distorted gravitational field.

The single unit of potential energy at the beginning would divide into more and more units of kinetic energy, until they stopped dividing and reached the Grid, like ball of rock falling down a mountain breaking into pieces. And that will be that.

CHAPTER TWENTY-SEVEN

A Digital Zoom

At the beginning of a sequence, the quadrangles are large, there are not many dividing lines, and little information; at the end there are lots of sub-pixels, and the maximum amount of information. The general pattern of the universe, from the point of view of a hypothetical Foundation or a resolving Frame which does not experience space and time, is that it is getting more dense, and increasing in mass.

When a Focal Point is zooming on the other hand, it is remaining at the same mass. Even though it is getting 'smaller,' because it has the same number of units of information at each level whilst the overall number increases, it is not getting any denser, and has a steady mass.

If a zooming Frame, which is steady in mass, zooms not by the average becoming one quarter smaller at each level, but instead say becomes one sixteenth its size, such that it maintains exactly the same proportions by the very finest measure, it would be as if it was reaching towards its own next level of information, or jumping a level.

Extending this idea, it would be possible that a Frame has within its own pattern, at every level it exists, all the possible future paths it may take, or divisions, such that each step has copies of all that may come in the sequence.

This can be thought of as its Digital Zoom, and it creates the possibility of there being echoes of a patterns own possibilities, within its own pattern.

The Focal Point becomes a Focal Range which spreads out along the zooming dimension, rather like before with diversity repeating down levels, but in this case as 'ripples of mimicry,' stretching out in front or indeed behind the Focal Point.

So far this just makes it a very diverse pattern, or a complex form of mass, with layers within itself that replicate its own construction. It is a huge leap in complexity, but it is not yet enough to create space and time from mass.

Patterns of energy which are zooming, or maintaining the same mass, whilst the rest of the pattern resolves, are 'objects' that have consistent mass in a universe of varying mass, but they do not exist in or 'experience' space and time.

This, as mentioned, requires something much more significant.

It requires the birth of life.

Nothing else in the universe experiences space and time like life.

This may sound rather bio-centric, but life is far more complex than any other form of mass, and experiences its own view of the world. Just think of its tiny DNA, our digitally encoded blueprints. No other mass has anything like DNA.

CHAPTER TWENTY-EIGHT
Shifting Focus

In order for life to come about, and space and time, as they all come together, we begin with patterns which have driven up their diversity, such that they are utilising all types of function, or dimension, both as a frame in x and y, and with ripples of its own zoom.

It has created two distinct dimensions, or ranges, from the combination of x and y. One is an increased range of resolution, the other an increased range of zoom.

The easiest way to represent the next big step which describes life, is to consider that a diverse Frame which has resolved and zoomed becomes synchronised with all the other possible patterns of what is commonly referred to as a 'multiverse'. It's steps or waves 'inwards and outwards' on x and y and 'upwards and downwards' digital zoom wise, enable it to be placed at the centre of all other possible very similar complex sequences, which may themselves be one or two or three or any number of steps before or behind its own sequence, spread around it in just the same way as it has spread out in its own dimensions.

All the multiple sequences can be considered as different Foundations laid out next to each other on the infinitely large television screen, or they can be considered as on top of each other, as different parts of the same Foundation.

In the first example, this opens up the possibility of a sequence jumping to a new Foundation, which in effect means it will end up with a different final outcome, or a different Grid.

In the second example, where all the ripples are part of the same Foundation (labelling the 'multi-verse' as part of our 'universe'), it gives a Frame the chance to 'jump' to a different possible sequence within its own Foundation, even though it will end up at the same location, and the same Grid.

Both are of course the same thing, with the idea that different Foundation Pixels are connected via their relative possibility, just the same as all the different sequences within the same Foundation, and is expressed by the sentiment in many a sci-fi film, that even if it is possible to change your fate, that itself may have always been your destiny.

In other words, where you draw the line between a multiverse, or a universe, or a monoverse, or perhaps even an omniverse, if not also a holiverse, is besides the point, given every multiverse is a universe, and every universe a multiverse, in infinite layers of possibility. The same is true with Foundation's, or sub-pixels, or Grid-pixels and such.

Every sequence is one of the infinite possibilities, and whichever way you outline it, the 'multiverse' gives a pattern the chance to cross over to a different pattern.

This level of complexity, with multiple dimensional possibilities all focused on the same Focal Point, can bring about the possibility that the Focal Point switch or move from pattern to pattern, and can be called a Shifting Focus, defined as something only life has.

In effect, by becoming the Focal Point of multiple universes, life separates itself from its own universe, or becomes a universe unto itself.

This doesn't at first sound very different to a pattern which can zoom and resolve, changing its possibilities with its own ripples, but when a non-living mass zooms, it can't resolve at the same time. Every time it divides, it either has to zoom or resolve, but can't do both, at each new layer.

When a living mass zooms, moving to a duplicate of itself with the same waves in all dimensions, it is able to both resolve and zoom at each level, simultaneously.

Simultaneous resolving and zooming can be thought of as the process by which the Focal Point is able to shift or jump from one pattern or sequence to another, switching universe, which is what you do whenever you move, or carry on existing from one moment to the next. It is also the process which filters information into what can be perceived as space and time.

CHAPTER TWENTY-NINE
An Inverse Clutch

In order to progress, as any sequence must, a pattern which is a 'non-shifting mass' either has to follow the exponentially quadrupling and 'expanding' resolution, or the exponentially decreasing and 'contracting' zoom.

It has to either zoom or resolve but can't do both.

In order for a pattern which is a shifting-mass to progress, it can do both at once, its Focal Point or Range jumping Foundations to do so.

When there is a possible pattern which creates a sequence constructed by both resolving and zooming simultaneously, one expanding and the other its inverse contraction, these cancel each other out such that it is as if the sequence is not dividing at all.

For a pattern or sequence which has a shifting focus, it can 'expand' exponentially by resolving, whilst 'contracting' exponentially by zooming, all whilst maintaining overall a steady amount of mass. Even though it is getting more complex by resolving, and zooming, it's mass or energy use does not accelerate but remains steady.

The quadrangles still divide, generating sequences which utilise all possible functions, but no matter how much a pattern zooms or resolves, a pattern whose sequence always zooms by exactly the right amount to cancel out any resolving, is seemingly going nowhere. Yet it is resolving, the information is dividing, and it is zooming, with every step it makes.

This is a bit like a hamster running on a spinning wheel. In order to get the wheel to integrate with the floor, and move forward, what is needed is the fourth dimension, often thought of as time, but which is more like a 'sampling machine' than another dimension.

With the gearbox metaphor, the ability for there to be movement, involving both space and time, is like introducing a clutch which works by being able to vary the size of the cog, or redefine the value of 'one'. That is no matter how much it resolves and zooms, because they can now be applied simultaneously at each layer, and one is the inverse of the other, their total combined can always be taken to equal one.

An inverse clutch can vary the timing of the whole mechanism, by contracting or expanding 'one,' such that it can also act like an accelerator, which enables more or less fuel to be consumed. More fuel means the functions and their inverse separate more, but by adjusting the size of the cog, or the definition of one, the progress remains the same plus one, despite the fact it has actually changed exponentially.

Although the new function can be thought of as the 'fourth dimensional type,' it is not 'time' as such, but being applied to the other dimensions acts like a filter, enabling the information to be separated into linear sequences of 's' and 't,' which from a suitably complex perspective with 'memory' enable the experience of space and time. On this dimension, the dimension which life travels on and which is made of 'four dimensions' (three gears and a clutched throttle), at every new level and for every unit in the pattern divided, application of the clutched throttle increases the proportion of its energy which is resolved, and decreases the proportion which is zoomed, whilst release of the clutched throttle increases the amount zoomed, and decreases the amount resolved. The pattern of life can therefore increase or decrease the amount resolved and zoomed, or the amount of 's' and 't' in each of its units, despite, subjectively at least, maintaining a steady processing rate of information, xy.

In this sense the application of a fourth dimension, or an inverse clutch, creates something more like a gimbal, rather than a hamster wheel, reacting instinctively to the waves and countering their motion so as to remain steady, or continue to move at a steady rate no matter the terrain. This enables the complex pattern of life to survive, keeping level in choppy waters, or increases the odds of it maintaining its integrity despite the likely onslaught of random information.

This metaphor does not however describe how it moves, or seems to move, rather than just respond in a complex integral way. If you don't believe in free will this is not a problem. We are all just phantoms of a gimbal rocking on the ocean, which acting like a turbine generates the holographic aberration that is life. The gimbal rocks automatically, with the phantom thinking it is steering.

If you believe in free will, or even just will, not only can we keep our balance extremely intricately, but we can choose to lean left or right as a whole, with the gimbal becoming more like a surfboard, or even a boat.

Either way, be it by luck, instinct, skill, or a combination of all three, in order for a life form to survive and maintain its existence by moving in our shared universe, it has to feather its throttled clutch by just the right amounts to maintain all the consistencies of its trajectory in existence for all its units of energy.

This is no simple feat, but then this is why it is a complex pattern.

CHAPTER THIRTY

Rippled Memory

The idea that all the different ripples of a multiverse are in fact part of a universe, is essentially the idea that the next layer of information in a sequence can be inside the 'present' layer, as with a fractal, so that when that divides in four, the next step in the sequence can be inside each of those new quadrangles, etc., ad infinitum, or up to the range of any universe, determined by the relative size of their Grid-pixels.

It could also be that each layer holds not just the next, but the one after that, and every other possible layer, with the ultimate possibility that in every unit of mass in our universe, there is enough room to hold the whole universe, past and present, at an exponentially decreasing scale. That there is an inverse universe in every unit of our universe.

In this sense, on our boat going through an ocean and with all the ripples correctly aligned, there can be waves of possible futures forward forming off the bow, and a wake of possible pasts left behind, by every single unit of energy or mass in a multiverse, the creation and disturbance of which enable movement to be created. Except there is no movement, as commonly conceived, only the pattern of disturbance.

The creation and disturbance of the waves, is the

movement. The boat, or the unit or collection of energy is not 'an illusion,' it does clearly 'exist,' as we exist, but it is the same as the ripples and the waves, and does not create them, or be moved by them. The whole system comes together.

The ripples, like pairs of mirrors with ever shrinking reflections, are patterns that were, as well as patterns that will be, and it is the variance in them that give anything the impression it is moving. The phenomenon can of course still be called 'moving,' but most realistically, if not conveniently, it should be considered 'information fluctuation'. Our conscious projections of existence are useful, and enable us to survive and be in our world, but they are distortions of perspective.

The ripples of possible pasts and future are possibilities of what will be, and of what has been, but the future remains entirely unwritten from our perspective, and is in fact completely random. The fact that we can ever predict anything is startling, crafting probabilities from possibilities, but then so is the fact we exist. Alongside these odds it is not so preposterous to suggest our memories are as random as our future, and it is essentially luck that we happen to remember what just happened, although of course that is what we are, a Focal Point of Chance.

Echoes of possibility provide connections, from a perspective when random responses are aligned. With a view of the full spectrum, the selection is because all possibilities can exist, but only some of them make sense.

This intersection of complex patterns therefore enables a Focal Point, and its Focal Range, to move, be it by apparent will or response.

It also enables a Focal Point to have the possibility of memory.

The echoes of itself spread around it, just the same as with a multiverse, bringing the possibility of moving along those echoes.

Non-sentient forms of mass don't 'move'. We may see them moving, but whilst their movement is a conjunction of their change and ours, it is more accurate to say that is a reflection of our movement, than it is theirs. For even if they are 'changing,' without the complex luck of 'memory,' there is no momentum, and the memory of mass moving through a universe is entirely our own.

It is perhaps semantic, but mass 'changes in density,' it doesn't 'move,' except from the perception of a living being.

CHAPTER THIRTY-ONE

Instantaneous Change

Even though our existence utilises all the various functions which information comes in, our consciousness maintains a steady linear rate, where every metre is the same, and every second is the same. This all has an obvious similarity with the process of 'calculus,' the mathematical formulation which converts a curving line on a graph into a string of tiny straight units, by generating the ratio of the tangent of the curve at any given point.

The central and paradoxical idea of calculus is that the rate of change can be calculated at any 'single point' on a line, as long as that line can be represented or programmed by an equation, which in essence means it is predictable and not random. It makes sense because the equation does predict what happens next, but it is paradoxical because the method calculates the rate of change at a single point, like telling how fast someone is running from a photograph.

To calculate 'the rate of change' at the smallest possible moment, and to use the method of calculus, you transform a given equation into a new form of the same equation, which if applied to the original tells you how fast it is changing at any point. All mathematical students know the method, such that you get a new equation which for any given value of x gives you the slope at that point, rather than a value of y as the original equation did.

The variables change from depicting the pattern of how two dimensions intersect over a large scale, creating a line for various quantities of x and y on a graph, into a ratio which depicts the smallest unit of one dimension, over the smallest unit of the other, at any single point along their varying rate of intersection.

The slope at a 'single point,' which is often explained to be 'infinitely' small, is comprehended as being the 'tangent of the curve,' which overlaps the curve for the briefest of moments, or the finest of touches, which is nonetheless long enough to determine a parallel.

This tangent of the curve can be projected as a straight line out towards infinity, enabling the accurate calculation of its exact ratio to any possible number of decimal places, as if magnifying the very smallest single unit of the curve onto a grander scale.

The really interesting aspect is that the smaller the units you look at, the more accurate the calculation, which results in the possibility of an apparently infinitely precise instrument, peering closer and closer at a 'single point' of the function which is 'infinitesimally small'.

It sounds convoluted, and is renowned for being difficult to conceptualise, because it is (just like space and time), but is essentially the ability to chop up what is continuous and 'curved' into units which are 'infinitely small and straight,' a bit like the analogue sound of a record or instrument being digitally sampled onto a cd. Except calculus is not a sampling of the continuum, like a cd, but more like a revelation of the digital recording lying much deeper than the analogue summary our brain takes. After all, a curve can be drawn through dots of data, summarising them, as well as dots of data be taken from a curve, summarising it (as a cd does), and our brain could be drawing a continuous pattern from what are actually far more detailed discrete samples.

The 'miraculous' part, or rather 'mysterious,' which appears to defy logic is that the closer you look, or the smaller the unit where tangent and curve overlap, the more accurate the measurement. This makes sense as far as it being 'pin-point' accuracy, but it is difficult to grasp in terms of actually making a measurement, because normally the smaller something is the harder it is to measure!

This is of course the nature of mathematics and its symmetry, not miracles.

The precision is not infinite, for nothing 'in' existence is, but the calculation can be made to the maximum possible magnification of our current universe, at its current rate, using mathematics as a microscope.

Where the ratio is an exact whole number, the ratio can be considered as an exact proportion, and predicted to stay that way forever, but where it is not it is seemingly possible to calculate closer and closer with more and more decimal places, as with the ratio pi or some other irrational number, apparently until the end of time.

In other words it is possible with the technique discovered to calculate the ratio at any point on a curve, to as much accuracy as you have time to consider.

The technique enables calculation at the highest level of accuracy in our universe focused onto a single moment of time, at a single location of space. It focuses the resolution of 'now,' onto a single unit of 'here'.

The reason it has a limit, or 'zeros in' on a limit, is because the very smallest margin any universe can measure is up to the edge of its own resolution. The 'inner edge' of our universe is obviously very very small compared to us and anything we deal with, but is also always getting smaller and smaller at its fastest rate as the universe divides (or expands), such that we could never catch it up, even if we could divide information as fast as it does, with the most powerful quantum computer theoretically possible (which is what the universe is).

This is why it seems like some cosmic tool unearthed by human thought, exposing as it does evidence of our arithmetical and finite universe, by digging towards the bottom layer using logic.

It works not just hypothetically but in the real world, and for instance revolutionised engineering (not least by guiding rockets and human beings to the Moon).

Whilst being able to consciously apply the method of calculus to complex equations is a challenge of high order, reserved for those with plenty of mathematical training, machines and our instincts seem to be able to do it automatically, when correctly programmed. For example sound which enters our ears is differentiated instantly by our brains from its natural exponential amplification, into a linear more steady graduation, such that we can deal with a greater range of noise.

This is the brain performing calculus, translating an exponential function into a linear one.

In the same way existence, or existing, is two exponential functions (and their derivative functions), one set the mirror of the other (and their many duplicate ripples), which sequenced together give the possibility of 'appearing to' walk forwards, or expand, at a linear pace.

Yet whether we are creating lots of tangents one after another from what is continuous and variable, or are zooming in on the tiny steps, nobody seems to know.

Comprehending conceptually what calculus actually is, or does mathematically, in terms of its 'infinitely small pinchers,' has always been left to our own imagination.

CHAPTER THIRTY-TWO

A Finite Base

The process of calculus touches upon not only the ratio of a tangent, and 'rate of change,' but also an irrational number, which like pi has unpredictable and apparently infinite decimal places of detail, known by its representative the letter 'e'.

Unlike pi however, which is of course the circumference of any circle divided by the length of its diameter, 'e' apparently attaches itself to nothing.

Referred to as the 'natural base,' or the 'natural logarithm,' because it's presence is visible in nature, it reveals itself not only in mathematical calculations such as calculus, but also such phenomenon as bacterial population growth, radioactive decay, or even compound interest rates at the bank, and is specifically linked to exponential functions.

This is because exponential functions, as mentioned, accelerate at an increasing rate. That is their rate of change, which calculus calculates, gets faster and faster like ever expanding feedback. In fact, because the rate at which all exponential functions accelerate is, like feedback, proportional to input, all exponential functions always maintain an acceleration rate, or rate of change, which is proportional to their own function, or the means by which the input, x, is converted to y, the output. That is the rate at which their acceleration accelerates, remains the same.

Dividing in two remains dividing in two and doesn't change, even though the expansion increases as the population does.

With $y = 2^x$, the exponential function of 'doubling,' the function accelerates with a 'feedback factor' not of 2, or 1, as you may expect, but 0.693... This is a bit like how open the exponential function has its throttle (or more accurately the steady rate at which the throttle is increasingly opened). The 'derivative' of $y = 2^x$, or its 'rate of change,' is always 0.693... times the value of y for whatever value of x is being input or 'read'.

$y = 4^x$, or the function which quadruples at every layer, accelerates with a factor of 1.386... That is its rate of change at any point is always 1.386... times it's value of y at the same point.

$y = e^x$, or the function which is multiplied by 2.718... at each layer (the value of 'e'), accelerates with a factor of exactly 1.000... That is its rate of change at any point is always the same as its value of y at the same point.

In these equations 2, 4, and 'e' are known as the 'base,' and when an exponential function has a base of 'e' (2.718...), which has been calculated to trillions of digits and is 'random' because there is no discernible sustained pattern within the sequence of numbers, the rate of increase is '100%' (with higher accuracy the more accurate your calculation of 'e'). This is why it is a special number, as if the feedback is 'in tune' with itself. It is also the function which is described as producing a rate of 'continuous' change, which is why it used in banks for interest rates, and predicts population growth of bacteria. If you have a bank interest rate of 5%, it is 5% of the growth rate of this function.

What this actually means and why this number rather than any other is difficult to conceptualise, even if the results and technique are practically very useful.

The reason it can be used for money and bacteria is because both of those are so numerous they can be thought of as 'continually' breeding or earning interest. This is the equivalent of a 'continuous' scale on the x and y axis, which means they can always get finer and finer. 'Being continuous' mathematically is as if there is no limit to the scale, or no smallest unit, which is sometimes quoted as 'infinitely small'. Yet this is impossible for anything that exists, for both bacteria and mathematics, both of which have a 'smallest unit,' which is of course 'one' unit. The single bacterium is obvious, the single number less so.

Calculus does not home in on the 'infinitely small,' it homes in on the 'one' single unit, the fabled 'atom,' or the 'small pebble' (which is what 'calculus' means). This represents the smallest unit in a system, the 'opposite,' or the 'inverse,' of the 'one' which represents 'everything'. It also 'homes in on e'.

So what is this 'smallest unit' of a system of information? What is the 'opposite' of the 'one' which represents everything, or the whole 'universe,' and what has it to do with the irrational number 'e'? In the dividing quadrangle metaphor the smallest 'one' is a unit of the Grid, at the opposite end of the scale compared to the 'largest one' that is the Foundation, and represents 'everything'.

Mathematically, or 'dimensionally,' it is only a tiny fraction of 'one' on the x-axis (quoted as 'infinitely small'), such that it seems to not make sense, but this is because 'one' on the x axis (or anything in the realm of our cognition), is not actually 'one,' but merely a marker to square up the 'scale of the x and y axis'. On the x-axis, the 'one' that is everything is whatever number is being read, with the act of taking a reading setting it to equal 'the total,' and the 'infinitely small' is the smallest possible fraction of this 'largest one' that has become the entire x-axis, which of course isn't infinitely small, but is actually the width of the x-axis at the time and place the reading is taken.

That is, whenever anyone or anything 'takes a reading,' which is akin to interacting with something else in the universe, they are lining themselves up with existence 'now,' and then beginning a calculation which taking time and space to generate, can get more and more accurate with more time and space, reaching maximum accuracy when they 'end'. Although the smallest unit in our universe is finite, it seems infinitely small because we can never calculate as fast as the universe. Even if you cut an object in half every moment of time until the end of time (which would be many times a second), although you would be getting very small very quickly, you would never catch up with the frontline of the universe, as it has been dividing since the Big Bang. Just as with chasing light around us, the very fast division below us is a wave we can't reach. The division moves so fast underneath us, we can't even think as fast. Our thoughts, and calculations, have the same rhythm of the universe, like everything in the universe, but travel at a much slower rate, skipping across existence like astronauts on the moon.

The 'scale' of our universe, to which we are tuned, is represented by the number 'e' (2.7182818284590452356… on and on for zillions of units and still counting…).

CHAPTER THIRTY-THREE

The Longest Square

The 'base' of a number, which is what we often consider as 'the number we count to before repeating,' is like the 'beat' of a sequence, in both space and time.

In the dividing quadrangle metaphor it is akin to the 'distance' between each layer of information, which being dependent on the function, was in their simplest or most harmonious and average forms 'doubling' and 'quadrupling'.

We count in Base 10, because we have ten fingers. It is the mathematical structure of our conscious cognition, laid upon the underlying mathematical structure of our physical existence.

A dimension in Base 10 is ten units long, compared to one unit wide, so to 'proceed' along this function as a sequence, as when we count, is like walking ten marks forward for one mark upwards, and then repeating, as if going up a set of shallow steps, or when averaged out a relatively shallow slope.

Our base ten numbers come in matching single forms, 0 to 9, which can then be put in columns, making calculation a whole lot easier than say roman numerals, and 'one' is just a marker to line up the tens, and one and ten are essentially the same thing, marks on the 'repeating' nature of a counting system.

It is also in possibility terms akin to a die with ten sides, where for every one step up there are ten possible steps forward, any number of which could be 'selected'. As the number ten has ten options, in base ten every number has 'ten options,' even number 'one'.

Base two on the other hand, or binary, is two units long for each unit up. It is also when sequenced the same as walking along two steps, for every one upwards, or in possibility terms like tossing a coin. In the dividing quadrangle metaphor it is also the same as a quadrangle having two states, such as being a square or a rectangle. It is the most simple sequence of information, which is why we have made machines that use binary code.

Binary is the 'foundational form' of information, and essentially the same as 'doubling'.

We use both binary and base ten to count, as well as other bases like twelve for the hours in one day, or three hundred and sixty five for the days in one year, although the hours in the day and the days in the year are of course not perfectly proportional like the purely mathematical examples of 'base two' and 'base ten,' which are like 'guidelines' of perfect symmetry.

Base is therefore not just a way to count, but also the way 'one' is divided. It is the way a dimension of information, which the axis x or y on a graph always represent, is chopped into numbers. The decimal system counts in tens, but divides in ten as well, and 'one' on an x axis, unless otherwise specified, is therefore the same as 'ten,' or 'one hundred,' or any multiple of ten. It is also the same as 'one tenth,' or 'one hundredth,' any of which can be set to 'one'. It sets the grid in the middle of an infinite scale, the same as our own perception, and unless we learn to 'think in single units,' or can somehow get a view of existence from the end of time, which given the complexity of consciousness will never happen, this will always be the case. Just like our physical senses, all our numerical systems rely on metaphoric summaries of the far more detailed actual numerical system on which we exist. Because our own scale of existence is so large compared to the mind boggling scale of the smallest aspects of our existence, and our expanding universe already has a very large base zillions of units long such that we exist in the middle of space and time, we can divide below one down to the 'infinitely small one' in any mathematical construct, yet never have a chance of reaching it before the end of time.

Calculus is extremely accurate, and it soon digs down to very fine detail far beyond our crude interactions, but it is never perfect. Even if a result is apparently 'perfectly proportional,' that proportionality can only ever be calculated to the level of detail of the universe. That is, if the rate of change is determined as an exact number, such as 'five,' it would be 5.0000... with another zero added for every next unit of existence, reaching maximum accuracy at the end of the universe, but never perfection. At the end of time it will be able to reach its longest, 'thinnest' most accurate state, and it's maximum quantity, but if the universe had been bigger the next unit may have not been a zero.

In this way, an unspecified x axis appears infinite in each direction, when actually it has limits, the same as a universe. An x-axis which represents everything that exists, can be taken as 'infinitely thin' from our perception, but it cannot be thinner than the thinnest aspect of existence at this moment, which if squeezed or stretched out into a 'single dimension' of binary code, or summarised by a single number, would be a line zillions of units long, or the number 'one' with a base of zillions of units. It is a line which is getting 'thinner and thinner' and 'longer and longer' with every unit of information added, zillions every moment on Earth alone, but it is still always a finite line. The width of this line, which is a measure of the width of the universe in a single dimension, is the smallest possible unit, and is the 'pebble' calculus homes in on.

If the division of length by width is harmonious, such that the length of a sequenced dimension is an exact multiple of its width, the smallest units generated by the division, or the pebbles which define the 'scale of the base,' would be perfect squares. If it were not harmonious, then the 'single unit' would be a slight rectangle. These dimensions, or collection of dimensions, which do not average around perfect proportionality, would harmonise around a different average, or a string of quadrangles. Their smallest 'single unit,' taken to have a value of 'one,' would when compared to a hypothetical 'perfect proportionality' or framework of pure mathematical symmetry, be slightly off the mark.

Calculus, which zooms in on the smallest unit (the 'width' of our universe in a single dimension compared to the length of a single unit), and harmonises over the irrational ratio or natural base 'e,' therefore reveals how far our existence is off the perfectly symmetrical average.

In a single dimension it is the base of a single unit of spacetime, but in two dimensions it is the area of a unit of spacetime.

It, or in two dimensions its square root, represents the average 'distance travelled' by energy at every division (the square root of 2 being the average distance travelled in a perfect square).

In fact, because we live in a world of averages, far above the chaos, everything in our universe or that which we interact with travels through spacetime in these proportions, and for any information which doesn't resonate on these proportions it is not part of our universe. In this way it is a bit like a filter which determines what information for us 'exists'.

It is the 'rate of change' for our universe in spacetime, as well as the rate at which spacetime is processed by anything in our universe, including ourselves. It sets the rate for '100% growth'.

The fact that it works in the our physical world, yet comes from pure mathematics, is a startling revelation of the fact that our existence is constructed from units of information.

CHAPTER THIRTY-FOUR

Energy and Mass

Energy comes in two main forms, that with mass and that without mass. Energy with mass is often just known as 'mass,' and energy without mass is often just known as 'energy'.

Of the energy without mass this also comes in two forms: that which is too 'fast' or 'big' to be in our range of mass, and that which is too 'slow' or 'small' to be in our range of mass.

An example of the first massless energy is 'light,' which is famous as the speed barrier to everything in our universe, and an example of the second is 'quantum energy,' which is the barrier from which small particles of mass emerge.

In the quadrangle metaphor light energy can be thought of as the lines which define the boundaries of the Foundation, as well as that do the dividing, and the quantum barrier can be thought of as the smallest units which make the Grid.

From our perspective with mass light has a fixed but unreachable speed, measuring the largest unit that can be travelled in any fixed time, and the quantum barrier is too small to reach, being a measurement of the smallest unit at the end of time, but actually light can be thought of as instant, or not moving at all, and the quantum barrier can be thought of as the 'current grid,' and thus also 'instant' and not moving at all. Even though we tend to think of light as the fastest speed of the universe, leaving us behind, and the quantum barrier as a seething cauldron far 'below our feet,' they can both be regarded as right in front of us, and a measure of our own 'speed' through spacetime, as well as everything that overlaps in our shared universe.

Massless energy defines the edges of our 'screen,' but since we are still moving in space and time, and not actually yet at the end of our universe, there is continuing 'exchange of information' between our range of mass now and the energy 'beyond' it, from both space and time left behind and yet to come, such that energy can become mass and mass become energy. What appear as the 'boundaries' of our world of energy with mass, and another interlacing or 'surrounding' world of energy without mass, generate the possibility of 'energy' changing from one form to the other and creating for us a sense of 'movement'.

These two main barriers which surround and pervade us, as if one is 'in front' or 'beyond,' and the other 'behind' or 'below,' even though everything is 'here and now,' seem like opposites, which they are in two dimensions, but they are also essentially the same, or 'become aligned,' when the information is in one dimension, which is similarly to the consistency of 'e,' an expression of the fixed rate of our universe's information processing (which being fixed defines it as a universe).

To consider how this is so, in the dividing quadrangle metaphor, the dividing lines of 'light' are by definition as thin as they possibly can be in order to exist, but at the moment of reaching the Grid (which from an unimaginable viewpoint outside of space or time they do instantly), they are as wide as the unit they are dividing, with their width equal to, give or take half a measure, the size of a Grid unit. The thinnest line in existence meets the smallest unit in existence and can divide no more, which is what all the information in one dimension is.

Because of the set condition that the Foundation can divide into the Grid, paired as they are outside of spacetime, they are in effect 'aligned,' as are all the possible sub-grids which could form in a sequence between, and in an order which is generated by a consistent dividing in two. Even with a 'shifting-focus' as complex as ours which can be thought of as jumping from one universe to another in a multiverse, and therefore from one Grid to another, all the possible 'sub-grids' of a sequence must form a link from Foundation to Grid, so that all the universe's align. Light speed, or more specifically its consistency from any perspective, is an expression of this alignment.

As with how 'e' is a measure of the possible distortion of a Grid unit from a perfect square, so light speed is a measure of the possible distortion between the ratio of the Foundation and the ratio's of all Grids. From our perspective, which moves through spacetime at a steady rate, and measures a consistently accurate and finite light speed, the sequence of sub-grids must maintain a constant proportion, aligned as they are with our Foundation. In this way, our multiverse is not 'distorted,' but 'in focus,' as it must be for a perspective to experience steady space and time, for if light speed varied in time, our experience of time would vary in time!

Both light, and the smallest units, are essentially the same edge to our universe, with a predetermined if random and unknown future convergence point, at the end of time and space. Yet it is us as a complex 'Focal Point' of information that are doing the processing, and generating a sense of 'now moving'. The speed of light, which can actually be of any 'value,' is the measure of our own movement, and its consistency is an expression of our consistent rate through spacetime. We are surfing the wave of the 'sub-grid' quantum barrier right below our feet, every moment able to 'disintegrate,' and when a beam of light streams from the sun or some long past distant star into your eye, where it 'becomes mass' by interacting with you and agitating your cells, it is the universe dividing and opening up new patterns now, 'right in front of you'. For even though it is often stated quite correctly that when looking at a star or distant galaxy you are looking at the past, because of the time taken for the light to travel to Earth, it is also true that from your perspective the information being processed by your brain is always happening now, and that it is actually two-dimensional information which gets 'translated' into a three dimensional 'universe' with space and time. Only a complex perspective inside filtering the possibilities experiences space and time and a finite 'speed of light,' and 'speed of time,' in time.

CHAPTER THIRTY-FIVE

Relativity Revisited

We can use our dividing quadrangle metaphor to imagine what happens when the pair of twins, or clocks, are separated as one goes off on a spaceship and the other stays on Earth.

Each twin can be thought of as a dividing Foundation, or as a Frame that develops within a dividing Foundation, or even a Focal Point within a Frame.

As with every unit of energy, they are the centre of an infoverse, which began at its own beginning, depending how it is defined. That is, this could be taken as the moment of their conception, or their first thought, or when they became an adult, or as a subset of a larger infoverse which began with the first life form, or the Big Bang, or the eternal multiverse…

Being complex patterns they can be imagined as circles travelling like ripples down the layers of information, from their own Foundation to their own Grid.

The twins can be thought of as overlapping in a shared sequence, possibly from an imagined third person perspective, or separately each as their own sequence in which the other is essentially an environmental part.

As with everything that exists both twins are constantly dividing, many times a second, and in this way they are expanding resolution wise, but they are also zooming, and so shrinking, such that despite processing so much information by dividing, they maintain the same size and shape.

They are never actually standing still, even though they appear to be, like running on a treadmill.

This direction in which we are all constantly travelling without noticing, where patterns resolve and zoom getting increasingly complex, is not just moving forward in time, or in space, it is the rate at which information is processed, and it is this which determines our conscious sense of moving through time.

Even though we think of the fourth dimension as 'time,' because it is the fourth dimension which brings the possibility of time, it is also the fourth dimension which brings the possibility of space, and so they are in origin not so different after all. They may seem different, with an obvious example being how we can return to the same space where we once were, but not the same time, yet this idea is a fallacy. You can never return to the same space, the same as you can never return to a previous time. We are always moving forward, and if you go back to where you were born, or even just walk across a room and then back to where you supposedly were, you are in a new space as well as a new time. The space where you were has gone, just like the time, and you return to a different universe, or a different version of the same universe, which has changed not least because of your moving, or having moved. Just by being, we are always moving, and there is no standing still, even asleep on the sofa. We are constantly flying through space and time, at a rate so fast it appears continuous, and in this way to go back to the same space would be to go back to the same time, the same as if anyone could return to the same time, they would also return to the same space.

If you could go back in time you would be young and

forget you had been old, and if you could go forward in time you would be old and have the memories of your life.

This doesn't happen. Your acceleration in space does not slow your speed in time, it accelerates everything else in time, like H.G. Wells Time Machine, as if the rest of the universe becomes detached from yours, whilst you slip into a different future across possible information streams.

This is because space and time are both created from the same information distribution, and it is our acceleration, that is our existence, which separates them into the space and time as we experience.

If we accelerate we get more space and less time, but thanks to our digital clutch we keep processing spacetime or information at the same linear rate.

When the rocket-twin accelerates, requiring an increase of energy flowing through them, they keep separating space and time more and more, but with the clutch every new level or unit of information is still equal to 'one'. It essentially acts like a gearbox with zillions of gears, resetting the ratio of s and t at each with every interaction back to equal x and y which was their original ratio, before expanding again, which is also reminiscent of a digital piston, pushing us along.

As they do this, the space that they are generating every moment is increasing, and the time they are generating compensates by getting much smaller, like squeezing a balloon. Their own space has expanded, they are indeed massive, and their own time, which is that of their universe, shrinks, even though they carry on ageing at the same rate.

Again, this must be because they are the kernel of their own universe, where space and time are balanced, or created.

This is where life exists, or is what life is.

If the rocket twin then chooses to keep accelerating, they would need to keep increasing their rate of fuel consumption, getting more and more massive as they consumed more and more of their universe.

In fact the point at which any accelerated mass reaches light speed would be the point it had used all its own universe as fuel, and akin to a single very strung out one dimensional line at maximum zoom that arrived at the Grid, or the end of time, instantly. Even if instead of burning fuel they had utilised the slingshot effect, they would have had to get slung around greater and greater mass, getting more and more massive themselves, until they were half the universe, and with one more sling become all of it, or perhaps shot out of it, catapulted beyond space and time like an ancient Egyptian pharaoh. The escape velocity of life, which we call death.

The fact that if anyone accelerates away from the Earth they fast forward their environment sounds quite crazy but is not illusion and is established fact, if usually described the other way round such that the accelerated in space slows in time.

In this sense, relativity supports the Many Worlds Interpretation, and rather than imagining that by accelerating into space we would return to a future Earth, it should be considered that by accelerating in space we would return to a different Earth. Whenever we move, we influence everyone we interact with, and we influence them by not interacting with them too, as we ourselves were not influenced by those we almost met. In this sense it is not so strange there is a different twin awaiting your return compared to the twin who might of been if you hadn't left and had influenced them, even if it is their age that has changed rather than their personality! By racing off on your spaceship you make them older, because there are billions of Earths, and billions of twins. If the twin happens to be a determinist that believes in one world on the other hand, there is as yet no explanation as to why their own acceleration should fast forward the Earth and all on it, but not themselves, and that this would happen to any single person that accelerates. Unless we all want to slip into the solipsistic and psychotic narcissism where everyone claims to be the centre of the universe, with all others mere reflected images or non-playable characters in a computer game, it must therefore be accepted we are all just one of our own possibilities. Every moment, because we are always moving, we are becoming a different version of ourselves, shifting through different possible universe's, where travelling through spacetime is like moving in a hall of mirrors, and you yourself are only one of a series of reflections. Nothing ever travels or moves through a universe, it travels to other universe's which are very similar, but changes in such a way that it appears they are moving across their 'own'.

From the perspective of the clock or the twin that doesn't move, or rather doesn't change its rate of acceleration, they of course don't see their twin change universe, but rather see them take energy from their shared universe whilst expanding in the space direction, whilst reducing in the time dimension, getting slower and slower. They will never see their twin consume their shared universe, because it would take them such a long time to do so. They see their twin accelerate in space, and get bigger whilst slowing in time, but they don't witness themselves fast forwarding in time, which is happening in a different universe, and with a different twin.

We are all changing every moment, the different versions of ourselves, and our universe's, lining up in such a way that our movement works, taking different paths through different universe's, which is also as if each of us is a different sequence through a multiverse that 'comes together'.

What keeps us all together is the fact that no matter how we move, and how everything else moves in our universe down to the finest units of energy, there are always other universes aligned or on the same frequency ready in a suitable position of possibility, spread as possible patterns like ripples and consisting of all the possible universe's which can make sense of our paths, all the versions of which are random, with ours by luck making sense.

CHAPTER THIRTY-SIX

Quantum Layers

If a twin in their lightship slows back down to Earth speed, they would get smaller as they did, and have aged less than those on Earth, whilst from their own view the Earth would get bigger, and be further in the future than linear time predicted; a different Earth. Even though it seems as if they slot back into their original rate of time, their own sense of the rate of time never changed.

An obvious question is then "Can they keep slowing down?".

Its not immediately obvious how the accelerated twin may continue slowing down after they reach Earth, which is itself travelling at over one hundred thousand kilometres an hour around the Sun, such that perhaps we imagine them getting off and staying put in space as the planet carries on, although of course the Sun and all the space surrounding it are orbiting the centre of the Milky Way at over eight hundred thousand kilometres an hour, so it's confusing to consider how it is possible to 'stop'.

One way is that they get cold and everything they are made of stops moving. Another is that they would actually have to carry on getting 'smaller,' not like a star by becoming squashed, but by losing mass like a diet which makes you shrink. Of course, for the complexity of a human, at some point this reaches a limit, and they would die from being under weight, or cold, but if a human could get down to the size of an atom (but magically keep their complexity and still have cognitive abilities), the amount of space they would be using would be really small, so their level of time would 'have to be' really high (via chance if a coincidentalist, or by responding gimbal wise if you are a determinist). This would mean each unit of their process would require a lot of time. If they then returned home from their journey to inner space, their twin that stayed still would be younger than they are, although again none would have noticed their own linear rate change. Their use of time will have changed, using relatively more in each unit of spacetime, but their experience of time will have remained the same.

Whilst they shrank their space would of course have got very small, approaching zero, and so actually what would happen is they would become less complex, or simplified, and thus dead. If they didn't, and kept shrinking, they would either reach the Grid, at the end of time, or the 'quantum barrier'.

Here, as with light speed, is again the boundary to massless energy (or energy beyond our range), but seemingly coming from the inside rather than the outside. No pattern with mass can go beyond this barrier, and so there is no possible direct information of what lies beyond it, but it is the other edge to the range of our existence.

'Speed zero' and the quantum barrier can be thought of as the same, but the barrier is itself receding, as the universe is expanding, or dividing. As with the boundary at light speed, the barrier is therefore moving, and always getting smaller at the fastest rate, such that any Focal Point of energy can never cross it, or even reach it, the same as no mass can reach the speed of light.

What they can do however is enter the 'quantum realm,' which is where, with the very small and simple, that quantum behaviour becomes apparent. For even though all information is quantum, and something as large as the Earth is in its construction quantum, it is highly unlikely for us to ever detect it as so, as it would have to not only stop moving through space, but all the units it is made of would have to not interact with each other for a detectable unit of time. It will in fact be doing this now, but at a frequency far beyond our range. The whole Earth and everything that makes it are always disappearing and reappearing, just never from our perspective.

In this way, if we could ourselves get really small, we would increase our odds of zero interaction with any other particles, and we would see the units of energy that make the Earth moving in steps rather than continuously. Beyond a bit of a non-definitive point, the odds of spotting the quantum nature of energy are significantly increased. The 'hum of interaction,' if it was to be played as audio, would suddenly have gaps. When mass is slow enough to hear the tiny gaps, or small enough so the gaps are relatively large, it is said to be displaying its quantum nature. This would be to have entered the quantum realm. Beyond the barrier further down, any unit of energy would be too small or slow to exist here and now, not even for the tiniest moment.

There is a lot of energy at these levels, coming and going rather than being consistently present as mass, but the phenomenon is naturally not easily spotted from an evolved living creature's perspective. It is impossible to interact with any energy that is beyond the quantum barrier, but it is possible to witness quantum behaviour in the quantum realm, although even this requires intricate technology, as was set up by very focused human beings with the dual slit experiment, where humanity built a tool which ripped open the fabric of spacetime, and peered at its threads.

Some say each particle goes through both slits at once, or that each particle interacts with the subsequent particles not yet sent and those sent before, but what actually happens is because nothing 'observes' it happening, they don't go through the slits at all. Because the particles interact with nothing else, not even themselves, there is no information regarding which route the particles take in this existence, which is the same as saying the routes don't exist in this existence. There is no option, if the existence of which they are a part happens to continue, but to fill in the blank with all the possibilities at once. By not actually having a path, it is as if all were taken.

This seems truly strange, for even if existence is random, it would be assumed the path of the particle be 'chosen' randomly, but to randomly choose one would be 'to make something up,' rather than just defaulting to the actual reality, which is that every possibility is always happening.

It is precisely because the particles do not interact with anything else as they 'travel across the box and through one of the slits,' that they do not 'travel across the box and through one of the slits'. Because they are not 'interacting' they are not 'in space,' and their energy is spread across all the time of their universe. From our view it is as if they now have a good chance of 'jumping' in space distances that are noticeable, to the boundaries of the experiment, but this was always the case; it's just in a normal environment the odds are far far lower. If no one or nothing interacts with another particle, if its environment can be 'slowed down,' it 'jumps' in space.

Energy that really is below the quantum barrier is everywhere at once, able to emerge anywhere instantly or in the future. Energy that is just above the barrier sometimes jumps and sometimes doesn't, appearing and disappearing in space. Energy that is far above the barrier appears to stay continuously in space, following solid paths, like planets. The experiment showed that beyond our limited range of vision a full spectrum of possibilities exists, all of which we don't see, and into which energy can 'disappear,' or 'appear,' apparently 'randomly'. This is the literal backdrop to our apparent probability, and peering at the border of our existence we witness reality and the illusional consistency of our world breaking down.

It is possible all the other possibilities drop away, leaving only that which exists or has existed in this world the one reality, but there is no reason why all the other possibilities wouldn't also exist. In conjunction with the clear evidence from time dilation, by which our own acceleration in space accelerates everybody else's rate through our time, it makes more sense that all the other possibilities in fact do exist. The layer of our existence, between two exponentially dividing dimensions, is just one of infinite layers.

Beyond the speed of light, or below the quantum layer, are realms just like ours, connected but forever beyond reach.

CHAPTER THIRTY-SEVEN

Freedom to Believe

The metaphoric considerations that have been discussed in this writing are not just an exercise in abstract thought but intended to be a realistic image of our existence, and how we are probably a randomly sampled finite distribution of information plucked from infinite possibility.

If this is actually the case and we are to attempt the construct of an honest summary of the human situation, it makes sense to begin with one glaring consequence of this belief, which is that we are unlikely to have any free will.

Regarding this there is a lot of confusion about what is actually being debated, but the crux of the argument is whether or not a being can make 'independent decisions'. The main problem is centred around the idea that when making a decision the process has to be based on previous experience, especially that of the moment just before, as with any chain of events. Whether you go for a walk depends on what sort of mood you are in, which may have been influenced by what you had for breakfast, or what you just read in the media, as well as countless other memories and the present condition. This is pretty obvious. The only other influence is who you are making the decision, which is determined by your genome, as of course in the nurture/ nature divide. Clearly you don't choose your own genome, but the jury is out on whether you can choose to ignore the depressing news or the rain clouds that make you want to stay at home, and in short change your mood by the use of your own will. Seemingly we can and often do influence ourselves, but this of course leads to the conundrum on whether the choice to override your mood was itself based on other previous experiences, such as learning a walk in the rain can be very rewarding if you just get out the door! This is how the debate becomes entwined with the idea of whether we have a part of us that is separate from the rest of existence, aloof as it were from all influence, which is a subject that has been greatly discussed but never resolved.

For those who don't believe in free will, and think it is only conceptual and has no place in physical existence, every thought we have and every decision we make is always based entirely on what we are or what has gone before, as there is nothing else, with our self-awareness essentially responses the same as our instinct. There is clearly a higher level of learning in humans but the idea of self-awareness being reflection is a misconception as all experience and response is always building on what went before. Some people then bring up that this must mean we live in a clockwork universe where everything could theoretically be predicted, but that isn't the case because there could also be random elements, or indeed it could be that everything is random. It does however mean that every choice we make is in a way made for us, by circumstance. That each of us is circumstance, including anything we may describe as our 'will'.

The contrary argument supporting the idea of free will being an actual thing that has physical existence claims that our self-awareness enables us to detach ourselves from the chain of cause and effect, rising above it as it were. The best argument in support of this is basically that the whole is more than the sum of its parts, in that we are more than the chemistry that makes us because if you analysed us at the molecular level you would not perceive we had minds, which clearly we do. A being emerges which can reflect, and be conscious of what it is doing, separately from the chain of cause and effect. In fact, for existentialists we ironically have no choice in the matter and being human with self-awareness are forced to make free decisions every moment!

It is difficult to say who is definitely right and who wrong, or if there could be some free will rather than it being absolute, but it does seem that regardless of how free you consider our will, and even if we are all random automatons, there is still the fact that human beings can be more or less independent as individuals, depending on their nature and their options.

This is obvious with the example of how the rich have more options than the poor, which gives them more freedom as individuals to make choices, even if it isn't always that simple of course, because if the poor person has more wisdom than the rich it could be argued they have more freedom, as in ability to have a free will. There is also freedom in having less responsibility, as there is freedom in not making a rationally considered decision but utilising your emotions or instincts instead and just going with the flow.

Setting aside the complex challenge of defining or maximising freedom, regardless of whether our progress in time is considered chains of physical cause and effect or coincidental information, there are clearly variable levels of what could still be labelled 'individual will'.

In order to clarify between two distinct ideas therefore, what is called 'free will' is here defined as the unique ability of a human being with self-awareness to detach from the chain of cause and effect and be able to take a position and decision that is separate from all previous influence.

What is called 'individual will' is defined as the ability or circumstance where an individual can make a decision rather than have it made for them by someone else. A maximisation of individual will would be when the responses or decisions that most influence the direction of an individual happen within that individual, even if just circumstance and without the need for free will.

It is a subtle difference but basically we are either waves on an ocean, or surfers on a wave, existing as individuals whether that be entirely by circumstance or influenced by our own movement.

CHAPTER THIRTY-EIGHT

Believing in Rationality

If the honest conclusion is that we are most likely a random snapshot of an infinite information distribution it has to be considered what actually believing this would mean for our way of life.

For a start not believing in free will makes you question whether anyone can be held responsible for who they are, and therefore whether we are ever worthy of praise or blame and justified in feeling pride or guilt.

If you believe everything is response and there is no free will at all, this means nobody is at fault for the way they behave, or misbehave, in an existentialist sense at least. Being talented is all down to luck, and being a horrible person or a criminal down to bad luck. If there were a God of this world, it would not punish sinners, or praise saints, but accept everything as it is, hoping to be a positive influence.

If you believe there is some free will but a lot of circumstance influencing how well we engage it, that means you believe individuals are more or less at fault depending on who they are and the context of whatever event took place. If there were a God of this world, it would vary the punishment depending on the situation, even if the same sin.

The educated should know better than the ignorant.

If you believe every moment all humans (or perhaps all adults) are responsible for their behaviour, and that in fact there is no way to escape our freedom, you of course believe we are always responsible for our actions, regardless of circumstances. If there were a God of this world, it would punish every sin depending on the sin, regardless of situation, as there are no excuses.

Similarly sometimes a whole nation can be held responsible for its actions, especially if democratic, but even if not (because the populace chooses to submit to their tyrant). Others believe only the leaders, or the rich and elite who make the decisions are to blame. In this way the honourable captain goes down with their ship, because they feel responsible.

Some blame those alive today for the actions of their ancestors, or feel guilt for the actions of their ancestors, and some think the slate should be wiped clean, and the sins of humanity past nothing to do with us here today, even whilst they live with the repercussions and reverberations which pass down the ages.

There is a lot happening in the blame game, and always has been, but if you look rationally at the evidence, you have to at least consider that everything which exists is randomly generated.

However this does not mean you do not respond to it. In desperate times the best response to a murderer for instance may well be to execute them, or punish them in a prison as a deterrent to others, if with sympathy for their bad luck rather than blame. In more sophisticated times a more rational if complicated and no doubt costly response would be to educate them in prison, with the hope of their release and becoming a valued citizen, or to form a society and education system where there is very little reason to commit crime, or even have any prisons. It would also be acknowledged that increasing the punishment of one individual in order to put the fear into potential others should itself be a crime, the same as plea bargaining.

Another consequence of believing everything is coincidence is that anything is possible, and so we can believe whatever we want, including that the Earth is flat, or that we are in an alien simulator, or that the Norse gods are in fact real. It is perhaps a wonderful liberation! In fact, that anything may be true, or at least seem true, is itself a bit of a modern truism, thanks to the recent explosion of available information, such that according to some we live in a 'post truth world'.

Except of course, we don't.

The fact that everything is random doesn't take away from our ability to distinguish between the more or less rational ideas, where rational beliefs have a better fit with the evidence that is observed.

That is although technically the Earth may be flat (and not because all information is two dimensional but because it is a disc in a three dimensional space), this according to evidence has a minuscule likelihood compared to the idea it has a spherical shape, the same as all the other planets and stars in space. Anyone that is honest has to believe this looking at the evidence, even though rationally there is still a minuscule chance it is flat, as believed by all earlier humans that walked its surface.

It is also much more likely that humans made it to The Moon in 1969 than it was all staged, even though technically either could be possible and somehow signals sent from The Moon were faked, fooling everyone including the competitors from rival nations, unless they too were part of the conspiracy.

Anything is possible after all.

Although nothing is ever certain, rationality clearly makes sense, so if you find yourself with a superstitious habit, like touching a bit of wood after testing fate, tell yourself it is highly likely it makes no difference. Ask yourself what sort of world this would be if looking at a new moon through glass really did bring you bad luck, or everyones prayers were answered.

These things don't happen, and miracles past and present are obviously far more likely acts of imagination than acts of God.

Tried and tested science clearly provides a far more detailed description of how our universe works, compared to religion or philosophy, but ironically this does not banish either. This is because ultimately science is also endless, just like philosophy or any form of knowledge, and so will never provide a complete answer. Well beyond the comprehensions of any single individual, let alone the average individual, it will carry on until the end of time uncovering more nuanced detail, diving into what becomes exponentially more complex, surrounded by what is beyond human comprehension, just like the physical universe. Although this is not necessarily a bad thing, because it means science can never be complete and can keep human beings busy for aeons, just as any good hobby should, it does also mean we can't hope for an objective truth of our reality to be discovered but have to make up our own. Even if there is a Theory of Everything which is 99.999% accurate, and very almost as certain as can be, it would still be a summary of what is known, and nobody can know anything about what is unknown. Ours may be a finite universe, but in order to be so its edges are round up or down and we can't ever uncover the fringes of detail which are to us infinitely small or infinitely large.

Which brings us back to philosophy, or rational scientific

religion, which is all we really have.

There is a need for a modern graspable summary, but the rationality of our ultimate ignorance doesn't seem like much of a belief system to get behind, especially if rationality is just the chance of being aligned with your random circumstance.

Many philosophers have pointed out there is no reason to believe in anything, and it is often the ultra-rational that are portrayed as cold hearted killers, destroying humanity as if we are robots.

But this is nonsense.

Psychopaths are not rational, or overly rational (and neither are robots), but rather they are lacking in empathy, as if they are not fully conscious.

To be rational is to be honest, not to be a psychopath.

So to be rational, and believe in rationality, is by this view (or these semantics), to believe in honesty, or more specifically self-honesty.

Honesty is very much something we can believe in.

CHAPTER THIRTY-NINE

A Rational Rubicon

Human society would be better off if it grows up and becomes truly civilised, which it obviously hasn't done yet given there are still wars, crime, and people dying of starvation etc. In order to get there we need to surpass a critical mass of general rationality in the population.

We are probably still nowhere near this threshold, but there is cause for hope because all honest rationalists are drawn to similar conclusions about some fundamental basics of life, and this is regardless of whether you believe in cause and effect or that everything is coincidence, and must be true for any self-aware species. In this way they are multiversal truths, unlike the so called laws of physics which although consistent in our universe would be different in other universes.

First and foremost of the multiversal truths for any self-aware life form is of course mortality. It seems obvious, but life is impossible without death, and death is the most difficult aspect of our existence to truly accept and be honest about. This is because the living being is a feed-back mechanism which is honed to survive. What appears to be a life force emerged billions of years ago from chemistry, but is better conceived as an increasingly more sophisticated means to maintain chemical stability. Of course chemical elements and compounds don't aim to maintain stability, they just exist for much longer if they are stable, but life seemingly does. It is a balancing act, walking on a rope that got thinner but tighter as the chemistry that defines it became more and more complex. Whether there is actually feedback, or if it is all just coincidence, the complexity of life demands a system of stabilisers in order to make sense, such that those which survive seemingly want to survive, in a way that establishes the will to survive as the prime instinct of biological evolution, and the essence of being alive.

In this sense the more immediate feeling for a living being is the fear of falling rather than the joy of jumping. Together they form a duality, or a polarisation of response, but panic is more natural than laughter, which is actually a release of panic, most obviously with hysteria. Like a safety valve, laughing is shedding our constant unconscious anxiety in a comfortable environment. Confident emotions and actions have their place and their validity, indeed they are essential to survival, but they are all based on denial of a more honest analysis of our situation, which is that we will inevitably and unavoidably fall. That is no matter how safe, life is and always will be in a fundamentally precarious position, which all honestly self-aware must acknowledge. Immortality is not the ability to keep living, it is the impossibility of dying, and in this sense there can be nothing alive that is immortal.

Some people say they are happy to die, but this is also impossible. What is possible is that you are able to feel happy by being able to deny your fear of dying, which is what happiness is.

A living being cannot be free of their fear of extinction until the moment they are actually dead and so anyone who says they do not fear death is lying, even those who charged gleefully into battle or committed suicide calmly at home. Their embrace of death is ironically the fear of dying, expressed as a fear of living. You kill yourself as you charge the enemy, to be free of you fear, and even if you feel you want to die your body will fight you to live, gasping with its last breath and still pumping your blood whilst it spills into the Earth.

We can rise above our mortal fear consciously, which is how we experience the likes of pleasure and humour, and love to be confident, striding and conquering, but it is always there and it can break through our veneer at any moment. It erupts to the surface like the lava beneath us, but we rarely even acknowledge it, let alone give ourselves the chance to comprehend it, as all is done to channel it away and bury it. Tiny cracks are exposed daily, for example in an anxious or aggressive thought, but if you actually believe you are going to die, say by having a near death experience, or find out you or someone you love are going to die sooner than expected, say by being diagnosed with a terminal disease, then you may realise the true source of our being. Then you can see the power death has over life.

In this way every human being is actually neurotic, even if some are more neurotic than others, (which is what the term fairly defines - those excessively neurotic). Of course being neurotic about not dying is understandable (as every living being is), but it also makes us humans highly susceptible to irrational beliefs (including arguably fearing dying). Because we are essentially in a state of panic at our own existence, as children we will believe almost anything, such that cartoons are real, or that perhaps we may have super powers, or there are ghosts in the cupboard. We supposedly grow out of this, learning to trust our environment, but with our rationality always a means of responding to our irrational foundations, the truth is we grow on our past, keeping it with us, and never out of it. Instead we learn to reflect on our emotional outburst, after the event. We learn to temper the child within, building our adult psychological structures on the shifting sands of our youth, forgetting that our ultimate hope was once to be immortal.

Any rational living being that is honest with their awareness must understand this fundamental insecurity of being alive. They will also have to conclude that even if there is no absolutely independent free will, thanks to the interconnectedness or randomness of existence, the centre of any self-aware being is with their own individuality, because of our self-centred conscious view of existence.

That is self-aware species must see or even construct the world inherently from their own perspective.

In fact we are so embedded into our personal view that if we are honest all we can be sure of is our own self. Our consciousness is the most complex phenomena in our existence, and there is no way of knowing whether it is the only phenomenon in existence. 'I think therefore I am' is the famous axiom and it may be that I am all there is, with everyone else an aberration of my mind! Of course you may protest and tell me you are the only one that exists, but there is no reason for me to believe it. Everything could be made up by my mind. I have no way to be certain of anything except my own existence, which is undeniable.

Believing the world a figment of your imagination may seem irrational, but actually makes perfect sense if we are honest. It is also rather psychopathic, disconnected from the feelings of others by believing they don't actually exist, or maybe psychotic, with the idea everyone else is in your head.

To think you are the centre of the universe is to be a child, but it is something we grow out of as we realise we share the world with other people, unless of course you are a narcissist, although even a narcissist doesn't usually believe they are literally the only person to exist.

It could be that each of us constructs everything, in our own universe, with each of us aberrations of each others mind. That is you may all be constructs of my mind, as I am a construct of yours.

That is we each exist, and overlap in this universe, but you are aberrations of my universe as I am an aberration of yours.

The logical conclusion is that there is only one being in existence, but one which has many viewpoints.

CHAPTER FORTY
Jungle Warfare

This dream of a world which is able to be rational (if only when it needs to be), is unfortunately undeniably challenging to bring about, given how self-centred most living beings naturally are. The biggest challenge is perhaps to civilise our instinct for power, be it over the environment, other people, or even just ourselves.

This is of course an extension of will, be it free or individual, although if everything is information and all cause and effect illusion then technically there is no power at all.

In the wild, where our instincts were crafted before the blossoming of self-awareness, access to power is quite literally vital. Every second we survive is an act of power, and the challenge that we each face just by continuing to exist extends into the environment and of course each other as we interact and make our way through the world, consuming energy in order to do so. The pursuit of power is not only what life is, but it is what has evolved, crafting biological systems as water shapes a river.

Because of this reality, some people still believe the 'law of the jungle' to be the only truth. That is they believe ultimately there is nothing but survival, with our awareness having no more purpose than as a means to extend our power. It has even been intelligently argued that if there is no objective justice or anything beyond death, as seems likely, that there is therefore no morality and nothing wrong as such with war or even in some circumstances genocide and wiping out your enemy, children and all, as we do with insects or bugs. With no one to judge only a fool is bound by any rules beyond those that are given by nature.

Evolution has indeed been built on the continuously competitive battle to live, with no quarter given fighting to death and hunting infants for tasty snacks, and no rationalist can deny this. The no-rules combat game from which we have come has enabled the arms race of survival to generate species and that includes ourselves, such that we are all essentially the product of utmost brutality.

Also undeniable is the fact that the intense competition of war, which some believe to be the ultimate sport of cunning and intelligence, has benefitted humanity significantly, to the extent we may ourselves still be in the jungle, or certainly never have left the Earth and made it into space, had it not been our tendency to fight. That is, spending significant amounts of effort and wealth trying to kill each other has stimulated scientific invention and therefore in ways been very positive for our progress.

However despite these facts, and with utmost respect for the courage of a soldier, we cannot just cast off our awareness so easily, and what it teaches us, even if circumstantial and outside its original remit! Humans are different to other animals, and even if the nihilists that argue that might is right are not wrong, because there is no right and wrong, they are also not right!

They are basically simple.

With the conflict of life that lies behind evolution, as with war, there aren't no rules, there are simple rules. For animals in the jungle, if something can be eaten it is good, if it can eat you it is bad. In war if it is the enemy they are bad, and if they are on your team they are good. Despite our tendency to round up or down each others personalities, summarising people as great or idiotic depending whether we happen to like them or whether they vote in elections the same way (often also reduced to a binary choice), human beings and human societies are a lot more complicated. As with fictional stories it may make life easier if we know who the goodies and the baddies are, but in reality everybody experiences internal conflict between the 'good' and the 'bad' within themselves. This doesn't mean some people aren't more bad or good than others, in general, it just means in the detail increasingly revealed by any advancing analysis human personality is obviously a lot more complicated, which makes everything more stressful and difficult because it's easier to make decisions or form opinions if everything is black and white.

This is why sexism or racism benefits the stupid; they can generalise their way to a deluded sense of certainty.

Even though nobody can argue against the universal right of a life form to fight for its own survival, or deny the glory that must have been winning a battle, and the bravery needed to hunt, killing and violence if not always wrong are however always expressions of ignorance. They can be justified, but the very act of using physical force as a means of communication and power to influence is obviously much more simple than using language to occur change from understanding. In this sense although nobody can judge murder to be objectively bad, we can judge it to be inherently stupid, or less advanced than more complex and sophisticated means to make your point or display your power.

Similarly, because morality requires self-awareness, and is complex, we can also judge a lack of morality to be stupid. We are able to realise that humans are capable of more sophisticated ways of interacting and creating subjective meaning. Although the wild and untamed arena's generate intricately detailed power displays, they are as if nothing when compared to what the human is capable. Our creativity is literally astronomical, and making robots to explore the cosmos is far more sophisticated than making machines in order to fight over who can be boss.

CHAPTER FORTY-ONE

Civilised Conflict

When it comes to power play amongst friends and colleagues, times of peace are not peaceful, they just have much more complex battle lines. Continuously analysing body language like posture and voice tone we categorise each other into levels of threat or trust, forming a network of information exchange only a very small percentage of which we are ever aware of. Reassuring and testing each other with our words and gestures, we thrust and parry to determine who is friend and who is foe, exchanging incredible detail with the subtle meanings behind a joke or facial expression. Anything life threatening would of course be top of our attention, but for the most part we are free to focus on the complexities of our personal interactions, during which our psyches play tug of war assessing the likely power distribution.

What was originally about who is the biggest and has first dibs on food or mating rights, and who backs off and accepts second place, has become who is the best at sport, or the funniest, or most popular, or has the most money, or the best children. It is about who has the most influence over others or the environment. Personality types help, and coming across as confident, but so does circumstance. Someone in a truck on a road has more power than someone on a bike, no matter how tough they are. Those who want to be strong like trucks, and powerful noisy motors and guns; it is no surprise. Driving fast cars is about feeling more powerful, the engine under the throttle and the steering in your control, whilst driving them dangerously is about challenging your own mortality, which makes you feel successful. Others realise their power is in their beauty and use it to attract positive attention. Those that are clever want to be the most intelligent. Quite a few compete over how intoxicated they can become, or not become, despite ingesting more than anyone else. It is a show of strength to be able to remain steadfast in the face of adversity.

The desire for power is an extension of our individuality, and it is understandable why we instinctively want power be it by improving our talents or building our muscles, nurture often feeding nature by enjoying what we are good at.

We fight for power with every act and every thought we make, because this is what living is.

We fight for power in every single interaction we have, even when retreating.

Those who want to be alphas are people that actively try to dominate others. They project the battle within themselves onto the world, garnishing a sense of success by feeling in charge.

Those who want to be betas have learnt not to fight in the world, avoiding confrontation and doing as they are told, although they still battle with themselves, fighting their own self-perceived weakness.

In the detail everybody is sometimes alpha sometimes beta, with the battle inside becoming the challenge to know the difference between defence and attack, and if there is any choice the suitable situation for either.

In some interactions who has the most power and is dominant and who has less power and is submissive is very obvious, particularly within any explicit hierarchies shaped by strict rules as within organisations like an army. Outside these strict hierarchies age is a general factor, especially in past societies, with a default respect or dependency to elders which is always ripe to be challenged.

In personal balancing relationships who is in charge should swing back and forth frequently during interaction, to the point that overall it is as if none are in charge, and none are submissive.

In a regular conversation it is usually the person talking that is holding the power baton, which is why teenagers often shout over each other, even more than the rest of us, desperate to make our mark. You can't let people dominate you and have to join the conversation, but equally you don't want to be arrogant and believe you have more rights than others, or be unable to submit and respect someone else's more knowledgeable opinion. For the rationalist it is obvious that to aim to dominate someone else as much as possible is as detrimental to maximising honest reflection among the species as is being too often dominated by someone else, or being obstinately individualistic, as in pointlessly rebellious beyond your developmental stage. It is obvious that the availability of all opinion maximises complex intelligence, as does the ability to listen and learn.

A rational self knows that in an honest society all the positives in life will be maximised, and in a closed society, with closed people, that all the negatives in life will be maximised. A rational person understands that everyone would benefit individually and communally by being open and honest.

This is actually quite simple to understand.

We don't have to become saints, or be generous with our

time, we just have to be honest. We have to consider that when someone does you wrong they do not know what they are doing, in that they have no choice. Of course neither do you, and so if you get angry back don't get angry at yourself for being angry! Whatever chains of cause and effect reverberate, rationally our response to any situation should be whatever is best for ourselves, this is true, but in order to be so we must also include our community, our larger society, and in fact the whole of humanity. We should actually include all life and the entire universe, if reaching for the greatest accuracy! Whatever you do, it should be what is best for the whole universe, from your own perspective. Complex decisions or opinions are difficult for everyone, but the agreeing on at least trying to be as honest as possible, with yourself and everyone else, is very simple.

CHAPTER FORTY-TWO

Power Pyramid

In groups of humans the default power distribution tends to the shape of a pyramid, as it does with many other primates, as well as predators and prey in general, with the more powerful making the minority at the top and the majority having less influence further down. This is the most simple power structure and is what happens when greed is unrestrained, which is largely as true for monetary democracies as much as tyrannies, because either way we are still ruled by our unconscious instincts. The financial system, though able to impressively coalesce all our various demands and supplies, has simply replaced the biological battle for energy competing to survive and breed with social need and the greed for wealth. As societies we are still all of us living in the jungle and have yet to find our way out. For the nihilists and the capitalists there is no way out, or at least they think without the pressure of survival and competition we won't generate freedom but instead weakness. They think without the fight there will be no power at all.

There is no doubt that having wealth detaches you from the pressures of survival, which is the essence of biological existence. So if humanity aim to end up living in a pleasure dome with every whim taken care of by robots there might be a lot of detachment. As a society we have to protect those that are weaker, like children, but we also have to learn to look after ourselves as individuals, and there is the likelihood if we are all protected and provided for, we will as a species become apathetic and pathetic. Searching for the weak spots in others and attacking them is an effective way to promote strength, as in the wild, but it is primitive. Every adult should ideally be able to look after themselves, but this does not mean we have to fight each other. As with our muscles, which grow in strength by being torn, we need pressure and confrontation in order to develop to optimum, and society shouldn't placate citizens as parents shouldn't placate their children, but neither war nor the stresses of potential poverty or death are still needed. As a civilised species we can choose to put challenges in our way on purpose, and should be encouraged to do so. We can choose to go to the gym, or choose to climb a mountain, without ropes if really wanted. You can set up personal situations in which to thrive on the turmoil, seeking it out as stimulation rather than the monotony of conformity.

Where we draw the line between observing ourself and dominating ourself, and observing the world and trying to influence the world, is part of our individuality, but the important part is to be honest about the interaction. Most of us hope society dominates killers and makes them submit to the rule of law, as we do those who drive dangerously on public roads, but we should also champion maximising individual freedom, whether that be in speech or action. Even though there is no such thing as being free, we should always be pushing for maximum individual freedom. This does not mean we have to be die hard individualists, it means that communal sacrifice should be voluntary as possible, such that we chose when we conform and be communal. It means that there should be a minimum amount of rules and a maximum amount of understanding. That we must be able to err on the side of risk rather than caution, if only by small amounts, because this is the means by which we advance. At its most extreme every one of us should have the power to destroy the world, and be trusted not to use it, but until then we should always be aiming to expand the boundaries.

The power structure has always been constantly challenged, which is the very pressure that generates it, but the ability to move past it requires individual independence. It requires the ability for every adult to be able to survive without any help from anyone else. Only with the true independence of every adult individual, can community and friendship and respect or love have any chance of being sincere. In a rational utopia there would be no power structure.

The possibility of our moving past the power pyramid that has governed all living beings since the first life form may be possible sooner than anticipated with the expected technology (the 'big three' being quantum AI computers, limitless energy, and genetic programming), but these developments could also ironically bring the possibility of a deeply entrenched power pyramid taking over the whole world, as if humanity are indeed, unsurprisingly, on a knife edge, with individual freedom on one side, and tyranny on the other. It seems that humanity will topple one way or the other eventually, and either form a society which is more like a gas, where every human can move about, or form a society which is more like a solid, where every human is firmly set in place.

It is perhaps a bit of an irrational dream to expect human beings to mature enough so we can all trust each other to be rational, and honest, without having laws to force beneficial social behaviour upon us, or perhaps even pacifying the genetic traits of future generations, but until then we can aim for every individual to at least have an equal privacy-transparency balance. That is the law should protect our anonymity equally in some aspects of our life, as well as enforce our honesty equally in others. This way we can encourage honesty whilst maintaining safety, but still have responsibility and statistically accurate behavioural models, as all important feedback by which we learn. Society as a whole should be totally transparent with all anonymous data. We should be able to see where money flows and pools with easily digestible information display, such that the gap between the wealthy and the less wealthy is a healthy inspirationally optimising distinction generating competitive inspiration rather than an unhealthy polarising resented divide. We should all be able to access and comprehend the primary symptoms of our societies both regionally and globally, minimising government but maximising individuality and multilayered communities.

CHAPTER FORTY-THREE
An Irrational Fantasy

The idea that we are living in a simulation is becoming more common and accepted as a possibility, but it is one programmed by chaos, not by aliens or gods.

In this sense there is a high chance all the information in our universe, and multiverse, is eternal.

This means every thought you ever had, and every interaction, lasts forever.

It also means that every action we make and every cell that twitches within us reverberates across the cosmos as we each communicate with the whole universe constantly.

With each of us the centre of our own universe, it could also be that when we die our universe ends.

It could be, when we die, we briefly comprehend the full complexity of our existence in a moment of eternity.

It could be that we all therefore die 'at the same time,' such that those that died before us die with us as well as all those after us in one great simultaneous transformation.

We could be a singular organism.

We could all be a part of a singular eternal being.

You can choose to believe you are on your own, or that we are all part of the same being, or that we are a multitude of beings, because it is likely that all these are true.

Beyond the immediate meaning we create together, whatever happens to humanity in the end, be it self-annihilation or rational liberation, with infinite possibility all the possible universe's can be organised by their complexity at the apex of which can be considered the 'perfect mean,' which metaphorically at least can be considered the 'infinite godhead,' or 'The Good'. It is a 'convergence of the ideal' which no human individual nor society can ever get close to, but we only need the knowledge of its inevitable existence in order to hope, and even strive, to keep heading whilst alive in its direction.